Polypropylene

Polypropylene
Structure, blends and composites

Volume 2 Copolymers and Blends

Edited by

J. Karger-Kocsis

Institute for Composite Materials Ltd, University of Kaiserslautern, Germany

and

Department of Plastics and Rubber, Faculty of Chemical Engineering, Technical University of Budapest, Hungary

SPRINGER-SCIENCE+BUSINESS MEDIA, B.V

First edition 1995

Originally published by Chapman & Hall in 1995
Softcover reprint of the hardcover 1st edition 1995

ISBN 978-94-010-4232-1 ISBN 978-94-011-0521-7 (eBook)
DOI 10.1007/978-94-011-0521-7

A catalogue record for this book is available from the British Library

Library of Congress Catalog Card Number: 94–68444

∞ Printed on permanent acid-free text paper, manufactured in accordance with ANSI/NISO Z39.48-1992 and ANSI/NISO Z39.48-1984 (Permanence of Paper).

Contents

Contributors

Dr. Z. Bartczak
Centre of Molecular and Macromolecular Studies
Polish Academy of Sciences
PL-90-363 Łódź
Sienkiewicza 112
Poland

Dr A. Y. Coran
Advanced Elastomers Systems
260 Springside Drive
PO Box 5584
Akron
Ohio 44334-0584
USA

Dr M. M. Dumoulin
National Research Council of Canada
Industrial Materials Institute
75 rue Mortagne
Boucherville
Québec
Canada J4B 6Y4

Prof. A. Galeski
Centre of Molecular and Macromolecular Studies
Polish Academy of Sciences
PL-90-363 Łódź
Sienkiewicza 112
Poland

Prof. P. Galli
Montecatini Tecnologie
Foro Buonaparte 31
I-20121 Milano
Italy

Dr J. C. Haylock
Himont Inc.
Research and Development Center
800 Greenbank Road
Wilmington
Delaware 19808
USA

Prof. V. N. Kuleznev
Moscow Institute of Fine Chemicals Technology
M. W. Lomonossow
Prospekt Vernadskii 86
117571 Moscow
Russia

Prof. E. Martuscelli
CNR Istituto di Ricerca e Tecnologia delle Materie Plastiche
Via Toiano 6
I-80072 Arco Felice (Napoli)
Italy

Dr R. P. Patel
Advanced Elastomers Systems
260 Springside Drive
PO Box 5584
Akron
Ohio 44334-0584
USA

Dr T. Simonazzi
Himont Italia
Centro Ricerche 'G. Natta'
P. le Privato G. Donegani 12
Cas. Post. 155
I-44100 Ferrara
Italy

Prof. L. A. Utracki
National Research Council of Canada
Industrial Materials Institute
75 rue Mortagne
Boucherville
Québec
Canada J4B 6Y4

Prof. S. V. Vlassov
Moscow Institute of Fine Chemicals Technology
M. W. Lomonossow
Prospekt Vernadskii 86
117571 Moscow
Russia

Preface

Crystalline polypropylene (PP) was invented in the early 1950s by independent groups in the United States and Europe. The commercial production of PP began in 1957 in the USA and in 1958 in Europe [1]. The reader will find different data, about this issue even in this book. On the other hand, there is no doubt that PP became the winner among the commodity of large-volume thermoplastics into the group of which polyethylene (PE), poly(vinyl-chloride) (PVC) and polystyrene (PS) belong, too. The mean consumption rate of PP was about 10% per year in the past, in given application fields this value was even higher. The forecast for the future trend of PP use is still quite optimistic.

Which are the contributing factors to this success? The key issue is related to the versatility of PP *per se*. This means that the structure and properties (including processability) of PP can be tailored to requirements. Modifications of PP can be performed in different ways: during the polymerization (e.g. production of syndiotactic homopolymers or copolymers with different comonomer content), in the reactor (reactor-blends), in compounding (e.g. manufacturing of filled and chopped fibre reinforced grades) or in further separate processing steps (e.g. wetting of glass mat by PP-melt, manufacturing of textile composite preforms).

PP, being a stereoregular polymer, exists in different tacticities and in various crystalline forms which affect its overall performance. PP exhibits many beneficial properties such as low density, relative high thermal stability, resistance to chemical attack, easy processing and recyclability. In addition, its mechanical property profile is very closely matched to that of the engineering thermoplastics. Some of the drawbacks of PP are its low temperature and impact resistance. This was the driving force of many R&D works which resulted in the commercialization of different PP copolymers and blends. This achievement was the next milestone in the history of PP and that is the reason why Volume 2 of this set is devoted to Copolymers and Blends. This volume contains comprehensive surveys about the nucleation and crystallization behaviour of the related systems. A concise chapter deals with the morphology development and its effects on the rheology and mechanical properties of PP-based alloys and blends. For readers interested in the latter issue with blends other than PP-based ones, Utracki's book [2] is recommended. It is intended further to demonstrate how the morphology, and thus the resulting mechanical properties, are influenced by the processing (drawing). Volume 2 finishes with a review on PP-based thermoplastic elastomers. Their manufacture by blending and dynamic vulcanization can be regarded as an adequate way for upgrading large-volume thermoplastics. Recently issued books in this field clearly indicate the reality of this issue [3–5]. I would like to emphasize here that several aspects surveyed in this volume had a strong influence on the state-of-the-art of polymer science. This statement holds especially when morphology control in copolymers and blends and factors of impact-modification are considered. Readers interested in the latter topic should refer to the excellent book by Michler [6].

This book is designed to serve as guidelines for property improvement, upgrading and engineering use of PP. It is for this reason that such important fields of PP application as textile fibres, packaging films and foils etc., are not covered. Nevertheless, I am sure that both the concept and realization of this book will fulfill the expectations of the reader and older books about PP will remain on the bookshelves of the libraries. Extensive referencing, included with most of the contributions to this volume, helps to deepen one's knowledge of the desired field. During editing I did not try to 'harmonize' the conclusions and opinions of the authors and my comments were restricted to a few footnotes. This is due to my belief that subjects of dispute must be settled by time, since that is the task of science.

Thanks are due to the contributors, who kindly agreed to cooperate on this 'venture' for their efficient work. I have also to acknowledge here the engaged work of the publisher Chapman & Hall, whose staff cut the time delay in the publication of this book, caused by the transfer from Elsevier Applied Science to the present publisher.

József Karger-Kocsis
Kaiserslautern and Budapest
A.m.D.g.

References

1. H.R. Sailors and J.P. Hogan (1981) History of polyolefins, *J. Macromol. Sci. Chem.*, **A15**, 1377–1402.
2. L.A. Utracki (1989) *Polymer Alloys and Blends*, Hanser Munich.
3. N.R. Legge, G. Holden and H.E. Schroeder (eds) (1987) *Thermoplastic Elastomers*, Hanser, Munich.
4. B.M. Walker and Ch.P. Rader (eds) (1988) *Handbook of Thermoplastic Elastomers*, 2nd edn, Van Nostrand Reinhold, N.Y.
5. S.K. De and A.K. Bhowmick (1990) *Thermoplastic Elastomers from Rubber–Plastic Blends*, Ellis Horwood, Chichester.
6. G.H. Michler (1992) *Kunststoff–Mikromechanik*, Hanser, Munich.

Contents for Volume 1

Contents for Volume 3

1

Manufacturing and properties of polypropylene copolymers

P. Galli, J.C. Haylock and T. Simonazzi

1.1 INTRODUCTION

The synthesis and production of polypropylene copolymers was a natural development in polyolefin technology after the initial discovery and industrial manufacture of the polypropylene homopolymer [1]. The production of random copolymers directly in the polymerization reactor significantly expanded the property envelope of this versatile polymer, giving lower melting resins with improved clarity, particularly in film form. Heterophasic copolymers with increased toughness and durability were developed by melt compounding blends of different preformed polyolefins. More recently, such polymer blends have been produced by polymerization of the monomers directly in the reactor. This is not only economically advantageous but has led to the tailoring of specific polymer structures with a unique balance of properties, enabling polyolefins to meet a wide range of industrial needs, from flexible packaging barrier films to stiff, impact resistant containers and engineered structural plastic applications.

1.2 ZIEGLER–NATTA CATALYSIS

At the heart of both random and heterophasic copolymer polypropylene production is the Ziegler–Natta catalyst system, developed from the initial discoveries of Professor G. Natta at Politecnico di Milano in 1954 [2–4]. The initial catalyst system used by Natta for the synthesis of polypropylene was based on Ziegler's catalyst system for the synthesis of polyethylene and consisted of $TiCl_4$ and an $Al(C_2H_5)_3$ activator. Unlike polyethylene,

Polypropylene: Structure, blends and composites. Edited by J. Karger-Kocsis.
Published in 1995 by Chapman & Hall, London. ISBN 0 412 58430 1

however, polypropylene can exist in a number of isomeric forms and Natta's initial studies gave a resin in which less than 40% of the polypropylene had the characteristics of a stereoregular polymer. The remaining polymer was atactic, with a low degree of structural uniformity and a rubbery consistency.

In order to develop a viable polymer with commercially acceptable properties, Natta's research was directed to a more stereospecific catalyst system. He soon realized that the polymer isotacticity was directly connected to the heterogeneity of the catalyst surface, and by employing solid, crystalline $TiCl_3$ obtained by reduction of $TiCl_4$, in polymerizations with $Al(C_2H_5)_3$ he obtained a higher percentage of isotactic product. The high isotacticity (about 90%) allowed a scaled-up industrial process to be developed by Montecatini, which contained a large section for the removal of catalyst residues that would affect product quality. In a remarkably short period of three years a scientific discovery in the laboratory had been transformed into a major new industrial plastic resin. The production process for this new polymer was, however, very complicated and expensive, requiring a large section for separation of the undesirable atactic fraction from the isotactic fraction in addition to the removal of catalyst residues. Because of its complexity, therefore, the process was not very versatile and was essentially restricted to homopolymer production.

Polypropylene research and development in the 40 years since Natta's initial discoveries and first generation industrial process has been concerned with improvements in both stereoregularity [5–9] and in the yields and efficiency of polymer production. This resulted, in 1975, in the discovery of highly efficient and stereospecific active $MgCl_2$-supported catalysts [10–12]. Such high yield, high isotacticity, superactive third generation catalysts are now widely used in today's commercial polypropylene production facilities. The polymer production process requires only three basic operations:

1. Polymerization.
2. Unreacted monomer flash and recycle.
3. Product finishing.

The need for atactic polymer and catalyst residue removal has been eliminated, resulting in a simple, vastly more reliable and versatile process.

1.3 THE MECHANISM OF ZIEGLER–NATTA CATALYSED POLYMERIZATION

The knowledge and understanding gained in the development of third generation, active $MgCl_2$-supported catalyst systems has made the polypropylene production process more versatile, allowing the synthesis and commercial production of novel polyolefins, random and heterophasic copolymers with properties never envisaged by the original discoverers of

Ziegler-Natta catalysed olefin polymerization. It is the understanding of supported catalyst architecture that is providing the major contribution to advances in Ziegler-Natta polymerization technology today.

The influence of the catalyst architecture is one of the most complex and difficult to control aspects of Ziegler–Natta catalysed polymerizations [13–15]. It is the physical and chemical structure of the active sites on the support that provide a three-dimensional shape that can be duplicated by the growing polymer particle. In a supported catalyst polymerization, when the monomers reach the catalyst particle, reaction occurs initially at the most accessible active sites. The polymer begins to grow, not only on the active sites on the surface, but also on those inside the crystal granule. The polymer chains grow outward from the surface, but also inward, causing the granule to expand progressively. It is this process that is particularly critical in the initial phases of polymerization. The mechanism of polymerization can vary from diffusion control to reaction control, and depends on the physical and mechanical properties built into the support and on the reaction medium [16–23]. Although kinetic studies of $TiCl_4/MgCl_2$ catalysts for polypropylene polymerization are inconclusive [24–28], it is well known that all these systems are characterized by the lack of a constant polymerization rate, and demonstrate the following general features:

1. A very high initial polymerization rate, followed by a rapid decay and a stationary state.
2. Maximum productivity at a reaction temperature around 60 °C.
3. An increase in the polymerization isotactic index with an increase in temperature up to 70 °C.
4. An inverse correlation between stereospecificity and productivity.
5. The possibility of reversibly varying catalytic activity and stereospecificity of the system by acting on the instantaneous composition of the co-catalyst mixture, in particular, the ratio between aluminium alkyl and donor.
6. Absence of monomer diffusive phenomena in the reaction mixture and throughout the polymer layer covering the catalyst particle.

Experimental results have been rationalized by assuming a chemical deactivation of some of the active centres and the presence of at least two types of species on the catalyst surface. These are: (i) isospecific polymerization centres which are unstable with time, and (ii) less specific polymerization centres which, in turn, are stable with time. The latter appear to be preferentially and reversibly poisoned by the outside donor.

In order for the polymer particle to provide true replication of the catalyst particle, the mechanical strength of the granule must be in balance with the catalyst polymerization activity. If reactivity is too high, an uncontrolled 'explosion' occurs, and the mechanical forces generated by the growing polymer chain fracture the granule into a fine polymer powder. If the mechanical strength of the catalyst particle is too great, a low level of reactivity occurs because the internal active sites cannot generate polymer due to lack of space. Good replication and high activity only occur when the mechanical strength and polymerization reactivity of a supported, third generation catalyst are well balanced [29, 30].

Figures 1.1–1.4 illustrate how a porous catalyst particle expands during prepolymerization, replicating its shape, the 'structure' held together by the intermingling of the polymer chains growing from the active polymerization sites spread throughout the catalyst particle [29]. Figure 1.1 shows a Ziegler–Natta polypropylene catalyst formed into the spherical shape characteristic of an activated $MgCl_2$ support. Figure 1.2 illustrates a cross-section of the growing particle during the initial stages of polymerization at a polymer/catalyst ratio of 0.1. The break up of the catalyst particle and the intermingling of the polymer chains growing from the polymerization sites can be seen. It is important to note here the first appearance of a continuous skin that surrounds the porous interior of the particle where the polymer chains are continuing to grow. It is this skin that provides mechanical strength to the growing particle. Figure 1.3 illustrates the growing particle at a later stage of polymerization when the polymer/catalyst ratio is of the order of 1.

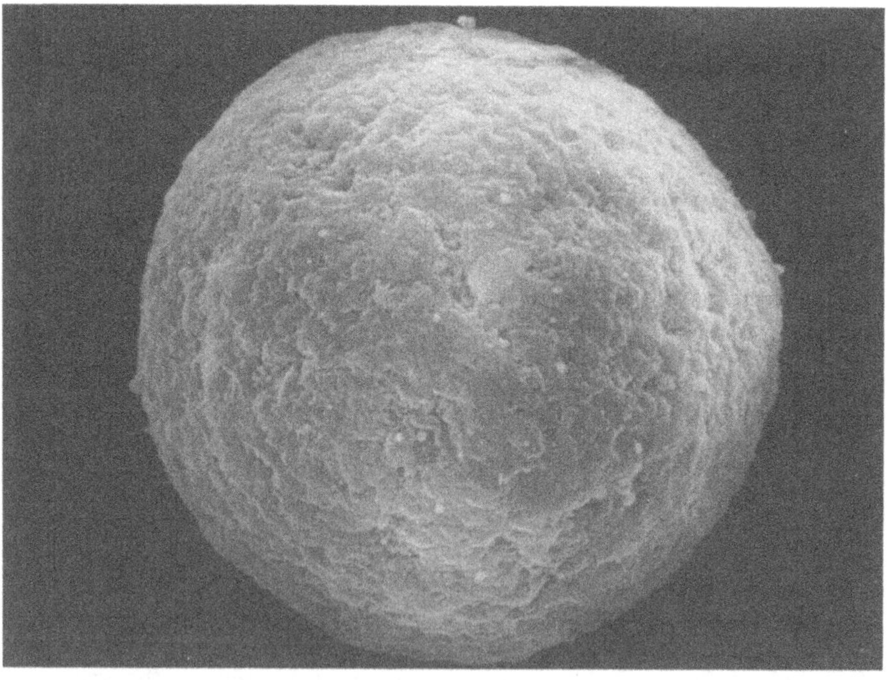

Figure 1.1 Active $MgCl_2$-supported Ziegler–Natta catalyst ($\times$ 1500).

Figure 1.2 Cross-section of polypropylene prepolymer. Polymer : catalyst ratio 1.1 : 1 (wt%) (× 3100).

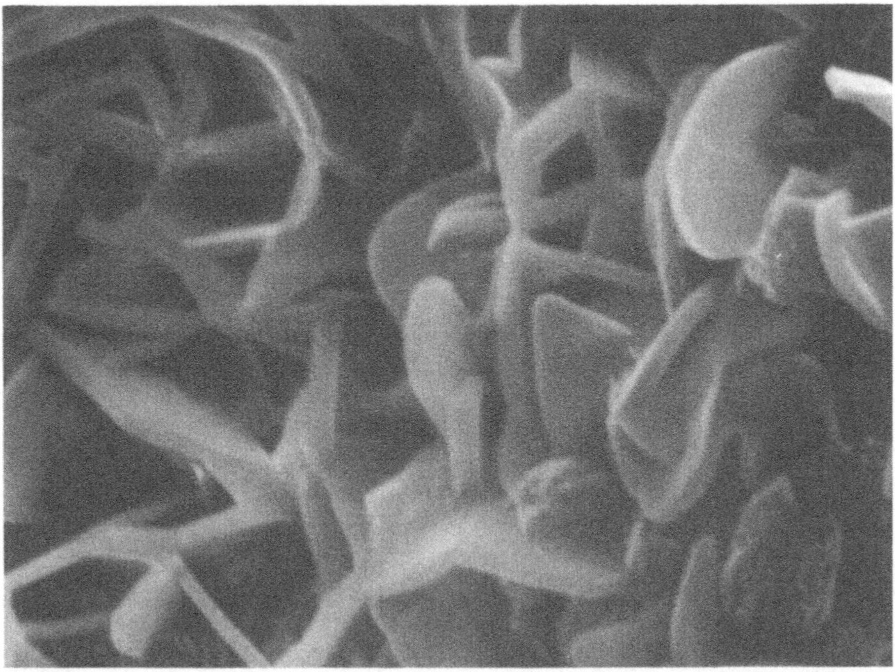

Figure 1.3 Cross-section of polypropylene prepolymer. Polymer : catalyst ratio 1 : 1 (× 5000).

Under appropriate conditions of catalyst preparation and polymerization, uniformly sized, compact polymer particles can be obtained, having the desired size and spherical form since they reproduce the original shape of the catalyst particle. Figure 1.4 illustrates a polypropylene particle at the end of the polymerization reaction. The duplication of the original spherical catalyst shape is obvious from this picture. Interestingly, it has been found in Himont's laboratories that only a few grams of polymer per gram of catalyst are needed to disintegrate the solid particle structure. Thus a large, active surface is exposed right from the very begining of polymerization. In the case of the super-active third generation catalysts, the replication factor (the ratio between the size of the polymer granule and that of the catalyst particle) can reach values of up to 40–50 in comparison to 7–10 with conventional catalysts, without forming fine particles.

The five basic, indispensable requirements for a super-active catalyst are:

1. High surface area.
2. High porosity with a large number of cracks evenly distributed throughout the mass of the particle.
3. High enough mechanical strength to withstand mechanical processing, but low enough to allow the forces developed by the growing polymer

Figure 1.4 Final shape of particles (× 22).

to break down the catalyst into the microscopic particles that remain entrapped and dispered in the expanding polymer.
4. Homogeneous distribution of the active centres.
5. Free access of the monomers to the innermost regions of the catalyst.

Manipulation of the catalyst architecture and control of mass and heat transfer during the polymerization allow relative changes in the polymerization rate in the interior of the granule compared to that on the surface. Diffusion limitations can be induced by controlling porosity and preferential distribution of active centres on the catalyst surface. The net result is that during polymerization the polymer granule can be varied from a relatively uniform density to a layered, open structure and even a hollow particle with a thick polymer skin. Figure 1.5 shows polymer particles formed under different polymerization conditions that illustrate this effect.

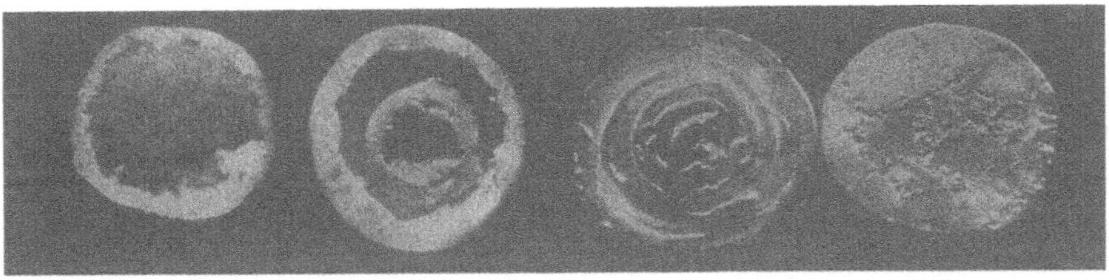

Figure 1.5 Cross-section of polypropylene particles formed under different reaction conditions.

It is the ability to controllably develop a growing solid, porous polymer particle that Himont/Montecatini has called 'reactor granule technology' [30–33]. The continually growing polymer particle can provide a solid reaction matrix, enclosed within which copolymer, terpolymer or multi-monomer random and heterophasic copolymer systems can be produced. The introduction of other monomers to the internal active sites during the polymerization process can produce a second or third different polymer within the original polymer granule, intimately dispersed among the pre-existing lattice structure of the initially generated polymeric matrix.

Control of the catalyst architecture and the reactor granule technology allows the freedom to conceive of and create new materials through the production of totally new polyolefin copolymers and alloys directly in the reactor [34]. Conventional third generation Ziegler–Natta catalysts, for example, limit the production of random polypropylene copolymers to relatively low levels of a second monomer. Low-melting random copolymers are difficult to produce in a conventional polypropylene industrial process because the 'sticky' nature of the resin causes fusion and blockage during the polymerization process. The porous, open structure of the reactor granule enables much higher levels of two or more monomers to be introduced, producing copolymers and terpolymers with unique properties.

Such low-melting copolymers can be encapsulated in a higher melting solid polymer matrix, effectively eliminating the problems of blockage in the polymerization reactor.

Since polymerization can take place within a solid polymer matrix, the mechanical containment aspects of the polymerization process became secondary. Bulk, gas phase and slurry processes are all equally adaptable for use with the catalyst system, and are chosen on the basis of economics, mechanical reliability and reaction control criteria in order to maximize reactivity and productivity.

This technique removes almost all of the previous process constraints on Ziegler–Natta polymerization, allowing reactor-made resins to be produced with properties no longer limited by mechanical considerations of the process.

1.4 THE POLYPROPYLENE MANUFACTURING PROCESS

Polypropylene was the first stereoregular polymer developed and its production was considered to be complicated and difficult to control. The first commercial processes used low yield catalysts and made homopolymer by reacting propylene with a solvent or diluent and a catalyst in a stirred reactor, which produced a resin that contained about 10% atactic polymer. The atactic resin was reduced to less than 5% by extraction with hot hexane or heptane. The catalyst residues were removed by washing with short chain alcohols and with water. To make random copolymers in a diluent plant a slurry of propylene and ethylene, solvent and catalyst are fed into a stirred reactor at 60–70 °C, at around 1 MPa [35].

Heterophasic copolymers require the addition of one or more secondary reactors where the homopolymer or random copolymer exiting from the first reactor is allowed to react with an ethylene–propylene mixture to produce an ethylene–propylene rubber as a second phase within the polymer matrix. The processes that used these early generation catalysts required the purification and recovery of solvents and the disposal of contaminated waters. The reactors frequently fouled and extruders often failed, resulting in a very limited range of copolymers that could be made.

The development of the high yield, high activity supported catalysts significantly changed the process economics and environmental impact and greatly expanded the range of possible copolymer structures that could be produced. It became possible to eliminate the diluent altogether by conducting the reaction in liquid propylene, thus creating what have become known as the 'liquid bulk' processes.

Some producers, however, have retrofitted their old slurry plants with the new high yield, high activity catalysts that can produce a resin with less than 1% atactic polymer, eliminating the need for resin purification and catalyst removal. The economics of that approach compared to the construction of a new liquid bulk process are in doubt, however. Although the de-ashing and atactic removal costs are eliminated the diluent must still be purified, and because high yield catalysts are more sensitive to poisoning, the diluent impurties must be significantly reduced.

The primary liquid–gas phase polypropylene manufacturing process used worldwide is based upon Himont's Spheripol process, developed in 1982 [36, 37]. The Spheripol process flow diagram is shown in Figure 1.6. Today 60% of the world's existing and announced polypropylene capacity operates under one or more Himont licenses, reflecting Himont/Montecatini's 40 years of commitment to research and development in Ziegler–Natta catalysed polymerization [35].

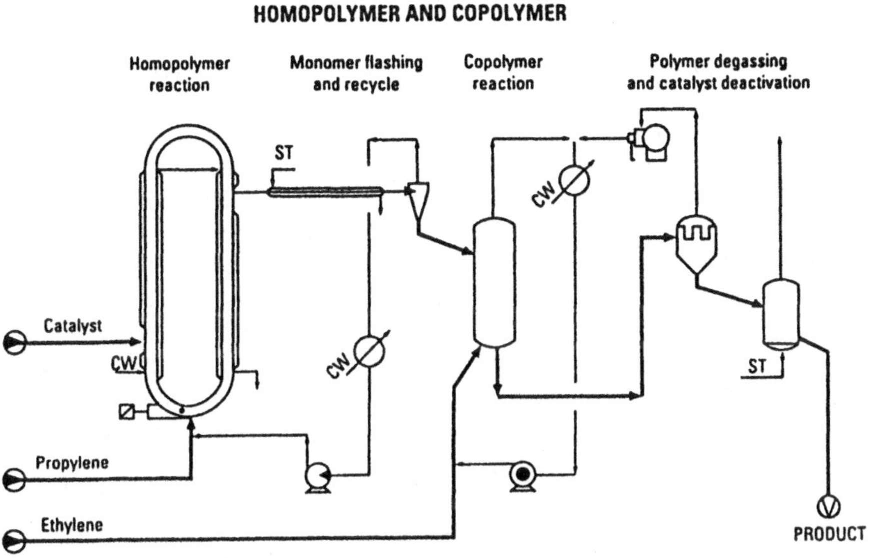

Figure 1.6 Spheripol process flow diagram.

In the Spheripol process, polymerization takes place in a loop reactor filled with liquid propylene. The three-component catalyst system, together with liquid propylene, and any other comonomers for random copolymer or terpolymer production and hydrogen for molecular weight control are continuously fed into a loop reactor in which polymerization takes place at 60–80 °C and 3.5–4 MPa.

A homogeneous mixture of polymer spheres in liquid propylene is circulated inside the loop reactor by means of an in-line axial circulation pump. The polymer is continuously discharged from the reactor to a separate cyclone. Unreacted monomers from the cyclone are then pumped back to the reactors. The final product is treated with live steam to deactivate the catalyst and to complete removal of unreacted monomer. This is followed by removal of moisture in the polymer to the low parts-per-million range. The resulting product is truly uniform because the highly turbulent liquid flow conditions in the loop reactor result in a very homogeneous reaction mixture. Process conditions are kept within strict limits, the temperature of the reactor being held within 0.1 °C by close control of a water-cooled jacket. Polymer particles as small as 0.3 mm or as large as 5 mm with identical properties and morphologies can be produced by the use of different catalysts. The content of the second or third monomer can be as high as 10%, although above 5% the comonomers tend not to be as randomly distributed, giving a 'blockiness' to the resulting random copolymers.

To make heterophasic, high toughness and impact copolymers in a liquid bulk process, the homopolymer or random copolymer particles exiting the first reactor are fed to a gas phase reactor operating at about 1.5 MPa and 60–90 °C. The solid, porous polymer particle entering the gas phase reactor has the accessible polymerization active sites in the catalyst dispersed throughout it, allowing copolymerization of propylene and other olefinic monomers such as ethylene into a polymer rubber. The rubber is heterophasically dispersed evenly in the solid polymer matrix, minimizing particle stickness and fouling of the reactor. The addition of a second gas phase reactor allows the synthesis of a third polymeric phase, such as polyethylene, inside the heterophasic copolymer matrix particle exiting from the first gas phase reactor. Thus, a complicated multiphase polymer structure with a homopolymer, copolymer or terpolymer matrix, uniformly dispersed within which are an ethylene–propylene copolymer rubber and a polyethylene homopolymer, can be produced directly in the reactor (Figure 1.6). The catalyst is designed to have polymerization kinetics regular enough to make possible propylene/ethylene–propylene sequential polymerization. Because the elastomeric phase grows on the crystals which previously originated the homopolymeric matrix phase, a morphological structure for the polymer alloy nearly approaching the ideal is obtained directly by synthesis in the reactor.

An alternative to the liquid bulk polypropylene process was developed in 1985 when Union Carbide adapted its gas phase polyethylene process to make polypropylene, using the super high activity catalyst from Shell Chemical Co. [38, 39]. A simplified diagram of this process is shown in Figure 1.7. In such a process, to make homopolymer or random copolymers, reactants and catalyst are charged to a fluidized bed reactor at 60–80 °C and 1.4–3.4 MPa.

It is claimed that mixing of propylene and ethylene in the gas results in a random copolymer in which ethylene is distributed very randomly. Ethylene contents of 6–8% in the copolymer are easily achieved in commercial production processes, and claims of up to 12% in pilot reactors without fouling have been made. Heterophasic, high impact copolymers are synthesized in a second reactor without the sticky, impact copolymers causing fouling problems.

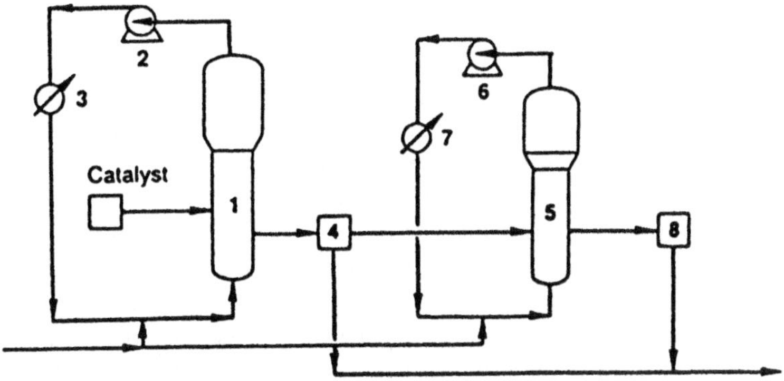

Figure 1.7 Unipol polypropylene process. 1, Reactor; 2, compressor; 3, heat exchanger; 4, discharge tank; 5, impact reactor; 6, compressor; 7, heat exchanger; 8, product discharge tank. Reproduced by permission from *Hydrocarbon Processing*, March 1991, p. 176.

Other companies have developed different polymerization processes including AMOCO [35] and BASF. For instance, the BASF process is gas phase and differs from Union Carbide's mainly in the reactor design [40]. BASF uses a vertical, helical stirrer at between 50 and 105 °C and 1.2–1.4 MPa. Reactors are one-seventh to one-tenth as big as the fluidized bed types and heterophasic impact copolymers require two identical reactors connected in series.

The polypropylene processes operating worldwide in 1985 were primarily slurry (66%), followed by bulk (21%) and gas phase (13%). However, the process choice for new polypropylene plants worldwide during the period 1977–1986 showed a clear trend away from slurry (38%) to bulk (38%) and gas phase (24%).

It is more difficult to give comparable figures for the relative growth of bulk and gas phase polypropylene processes since 1986 since the dominant process worldwide, the Spheripol process, is really a hybrid process having liquid-bulk loop reactors and gas phase reactors. The success of this approach however, can be judged by the fact that by 1995 approximately 50 Spheripol process plants will be operating in 25 countries, with a combined capacity of over 5–6 million tons per year. Furthermore, by 1995, more than 80% of the world's forecasted polypropylene capacity may operate under one or more Himont licenses.

1.5 RANDOM COPOLYMERS: STRUCTURE AND PROPERTIES

'Random' ethylene–propylene copolymers, with a low ethylene content, represent an interesting class of semicrystalline materials [34, 41]. Due to the insertion of a second monomer in the polypropylene-based molecular chain, these copolymers have a lower melting temperature than the homopolymer. The degree of melting point depression depends on the total ethylene content and its distribution along the polypropylene chain. Upon slow cooling from the melt the resulting polymer crystallizes into a mixture of α- and γ-crystals, and it has been demonstrated that the γ-form content increases with an increase in the ethylene content [42–44].

Guidetti *et al.* [45] found a linear correlation between the amount of crystalline γ-form and the number of short isotactic polypropylene segments. These results support the assignment that the development of the γ-form is determined by the interruption of isotactic sequences as the comonomer is inserted in the polymer chains. ^{13}C nuclear magnetic resonance (NMR) analysis has shown that no inversions exist is propylene units and that the polypropylene sequences, separated from ethylenic groups, show a high degree of stereoregularity. It is worth noting that the fraction of the triad PEP and EPP (P, propylene; E, ethylene) grow linearly with an increase in C_2 content. However, all trials with two ethylene units 'EEP' also grow. The amount of crystallinity due to the γ-form and the melting temperature depend not only on the total ethylene content, but also on the distribution of ethylene along the chain.

These results strongly suggest that the molecular microstructure of the random copolymer plays an important role in determining the properties of these very interesting materials. Thus, it can be concluded that most of the important physical characteristics of a random copolymer may not only be related to the overall ethylene comonomer content, but may be more accurately related to the concentration of a specific type of chemical defect. The ^{13}C-NMR technique has proven to be a useful tool in identifying the type and quantifying the content of such defects. It should also be pointed out, however, that the degree to which chemical defects affect random copolymer properties strongly depends on the nature of the defects. It can be seen that the defects characterized by the sequence $—CH_2—CH_2—CH(CH_3)—CH_2—CH_2—CH_2—$ (EPE triad) seem to have a greater influence than defects characterized by the sequence $—CH(CH_3)—CH_2—CH_2—CH_2—CH(CH_3)—CH_2—$ (PEP triad) and the EPP triad $—CH_2—CH_2—CH(CH_3)CH_2CH(CH_3)—CH_2—$.

The complexity of the structure of a random ethylene–propylene copolymer is easily demonstrated by proper annealing of the polymer. In Figure 1.8 a differential scanning calorimetry (DSC) thermogram is shown of an ethylene–propylene copolymer before and after stepwise annealing [46]. The increased resolving power of this technique permits measurement of the ethylene content distribution through the specific crystallization behaviour of the polypropylene segments separated by ethylenic units.

The incorporation of other olefinic units, such as butene, can give polypropylene copolymer interesting, commercially useful properties. Random terpolymers with ethylene and butene-1 comonomers give a higher

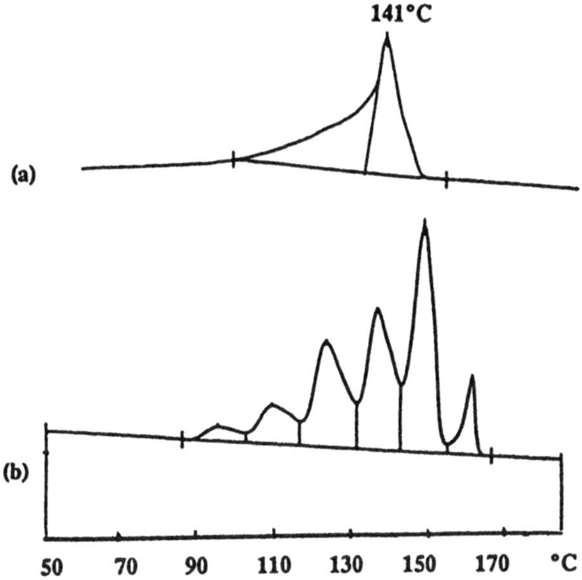

Figure 1.8 DSC thermogram of ethylene–propylene copolymer (a) before annealing and (b) after annealing.

percentage of the γ-form at the same comonomer molar content, resulting in a lower melting point compared to similar bipolymer [47]. An interesting random copolymer has recently been reported, obtained by the copolymerization of polypropylene and 1,3-butadiene [32, 48]. The saturated hydrocarbon structure of polypropylene and its ethylene and butene-1 copolymers have no polarity and therefore little chemical reactivity.

The use of 1,3-butadiene and the consequent unsaturation pendant to the polymer backbone offer the unique potential to synthesize new product through the reactivity of the carbon–carbon double bond or that of the corresponding allylic hydrogen. The use of super-active supported $MgCl_2$ catalyst systems for the copolymerization of propylene and 1,3-butadiene in liquid propylene produced random copolymers with different microstructures depending on reaction conditions and catalyst composition. The structure of the butadiene units can be controlled by either changing the polymerization temperature or the aluminium alkyl-electron donor ratio, resulting in favour of 1-2 addition of butadiene. This appears to be unique in the field of Ziegler–Natta catalysed polymerization since only copolymers characterized by a predominantly 1-4 addition of the butadiene units have previously been reported. Through reactivity typical of double bonds and allylic hydrogens, the unsaturated polypropylene copolymers can give a range of modified products that make it possible for polypropylene copolymers to compete in a wider field of applications requiring such properties as adhesion, cross-linking, covulcanization as well as in novel alloys and blends.

1.6 HETEROPHASIC COPOLYMERS: STRUCTURE AND PROPERTIES

Heterophasic polypropylene copolymers are tough, high impact plastics where polypropylene is the continuous phase and an elastomeric phase (usually an ethylene–propylene rubber) is uniformly dispersed within the matrix [49–53]. Before the discovery and commercial exploitation of the high activity, high yield $MgCl_2$-supported catalysts, such heterophasic copolymers were essentially made by melt blending of the preformed polymeric structures in an extruder. There were, however, significant limitations on the properties of polymers that could be blended in this way. For instance, too great a difference in the melt viscosities of the various components of the blend prevented the formation of the microscopic domains of low modulus ethylene–propylene rubber particles. An optimum size is necessary to deconcentrate stress during impact which, according to classic rubber toughening theory, acts to initiate delocalized energy absorption and reduce catastrophic brittle failure [54]. Figure 1.9 is a cross-section of such a heterophasic polypropylene copolymer where the ethylene–propylene rubber particles can clearly be seen uniformly dispersed in the polypropylene matrix.

To make the optimum heterophasic polymer structures directly in the reactor [55], it is important to tailor the catalyst not only for the

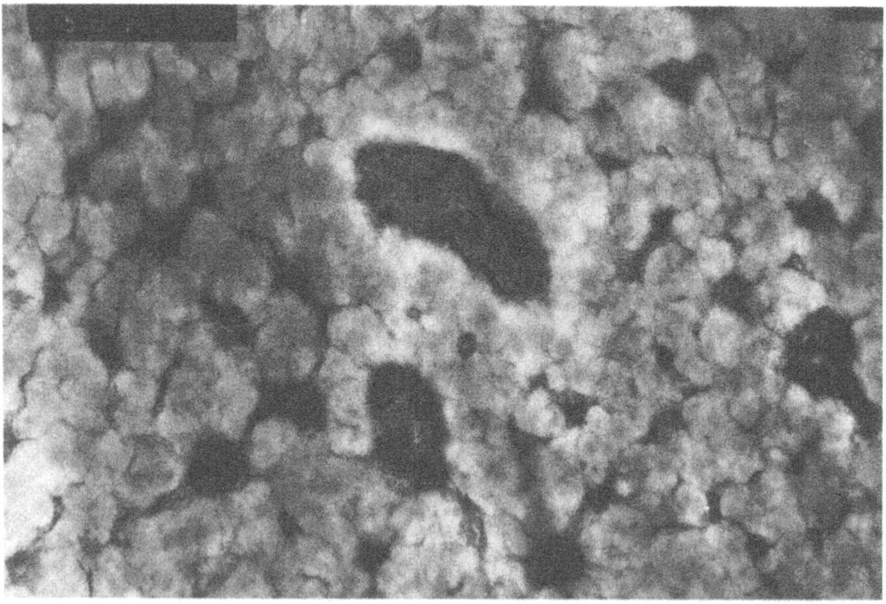

Figure 1.9 Cross-section of polypropylene–ethylene–propylene rubber heterophasic copolymer (× 11 800).

production of the ideal polymeric chemical structure but also for the location of the various polymeric phases. The rubbery phase must be homogeneously dispersed and its size controlled in order for the heterophasic polymer to achieve an optimal stiffness-to-impact balance. The catalyst must be capable of producing a homopolymeric phase with high isotacticity and then to polymerize the desired elastomeric material with a high degree of randomness, uniformly dispersed within the matrix.

Not only is the morphology of the catalyst system important, but so also is its kinetic behaviour. If, at the end of the homopolymerization, matrix stage activity is not sufficiently high, as was the case with early generation Ziegler–Natta catalysts, the system had to be reactivated by adding fresh aluminium alkyl. This, however, can cause difficulties in controlling the growth of the rubber fraction in two ways [29]. Firstly, the newly generated polymerization centres are not homogeneously distributed throughout the polymer granule (they are mainly located on the surface) and among the various polymer particles. This occurs because of the difficulty in permeating the polymer granule with fresh aluminium alkyl and uniformly spreading it over the different particles. This results in a corresponding variation in the distribution of the rubber since the overall system reactivity depends on the actual aluminum: donor ratio.

Secondly, the non-homogeneous concentration of the aluminium alkyl results in other drawbacks such as undesired and uncontrolled changes in the nature of the rubber fraction, since it is known that the aluminum: donor ratio also affects both copolymer composition and molecular weight. Similar effects could be caused by 'hot spots' arising from the high reactivity associated with the occurrence of localized high aluminum: donor ratios.

Today, with the high yield, high activity catalysts based on Ti-salts supported on active $MgCl_2$ and suitably modified with the appropriate Lewis base donor, it is possible to make directly in the reactor a highly crystalline polypropylene matrix with isotactic index values close to 99% within which is uniformly distributed an elastomeric phase. Because the elastomeric phase grows on the crystal faces which previously originated the homopolymeric phase, a morphological structure nearly approaching the ideal is obtained directly in the reactor [56]. Beyond the obvious advantages in energy savings by eliminating the need for melt compounding and extrusion, direct reactor synthesis makes it possible to obtain specific elastomeric phase structures that improve the phase compatibility and therefore the property balance of the final product. The amount of elastomeric phase, which continues to form between the polymeric matrix interstices, increases as the copolymerization process proceeds.

The performance of the high yield, high activity $MgCl_2$-supported catalysts has so improved control of polymerization of the rubbery phase that it is now possible to obtain polymers which can no longer be considered as just highly impact-resistant polypropylene, but rather as 'free flowing' resins with elastomeric properties and spherical morphologies. Design of the specific catalyst architecture and chemical composition as well as the polymerization operating conditions determines, through the polymer's growth mechanism, the control of the product morphology. This allows an easier processibility, a wider monomer compositional range and the appropriate distribution of the different phases inside each single polymer particle. The concept of the reactor granule technology, mentioned earlier, involves a controlled polymerization process taking place in each resin particle or granule according to the diffusion and kinetics phenomena, related to the structure of the selected catalyst. It produces a growing, spherical granule that provides a porous reaction bed within which other monomers can be introduced and polymerized to form a polyolefin alloy. Figure 1.10 shows the reactor granule technology, illustrating the porous, solid nature of the resin particle during the production of a polypropylene heterophasic alloy.

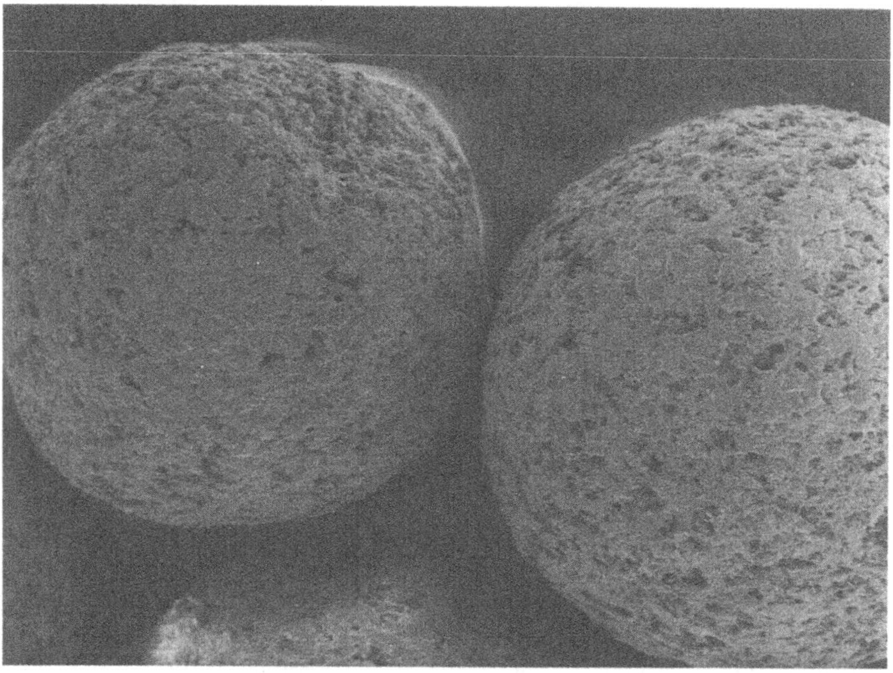

Figure 1.10 The reactor granule: a porous polypropylene heterophasic alloy ($\times 22$).

The reactor granule technology allows high rubber-containing blends and alloys to be made directly in the reactor. Control of the chemical structure of both the matrix and the elastomeric phase, as well as their individual molecular weights and molecular weight distribution, enables compatibility at the interface to be controlled for the production of optimum impact properties over a wide range of temperatures. The morphology of the heterophasic resin can be varied from the typical discrete phases shown in Figure 1.9 to an interpenetrating polymeric phase lattice network, as shown in Figure 1.11. Such heterophasic polymer morphologies provide significantly different resin properties, particularly in impact-modulus balance and tear strength of films.

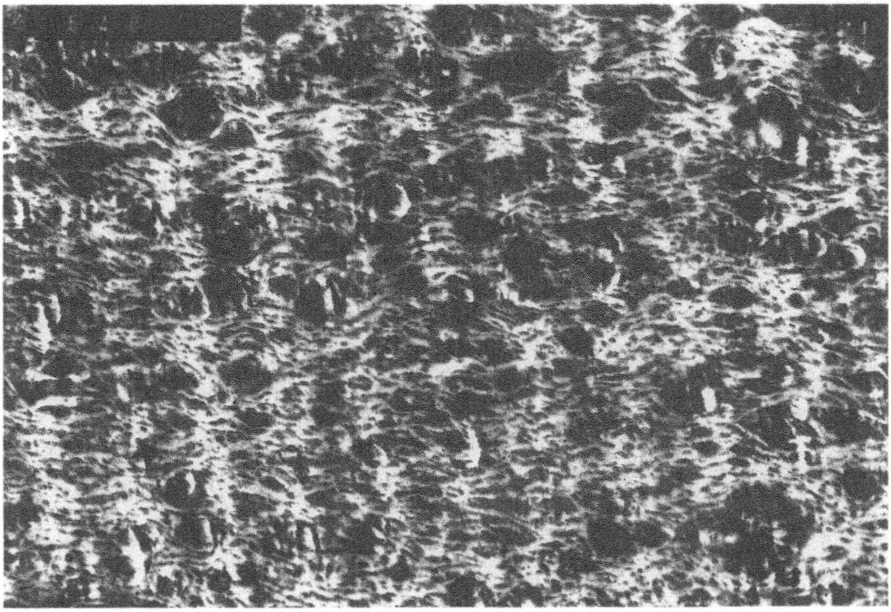

Figure 1.11 Reactor granule technology: interpenetration polymer lattice network structure of polypropylene heterophasic copolymer (× 10 000).

This technology is not limited to a two-component heterophasic system, however: a third or even fourth phase can also be introduced. Figure 1.12 shows a reactor-made polypropylene ethylene–propylene rubber heterophasic copolymer in which polyethylene has been incorporated to improve stress whitening on impact of a moulded part. The polyethylene phase has grown on the same crystal sites that gave rise to the ethylene–propylene rubber phase and is therefore inserted where it is required, on the surfaces and towards the interior of the elastomeric phase.

1.7 POLYPROPYLENE COPOLYMER APPLICATIONS

The new catalyst structures and the reactor granule technology allow the freedom to conceive of and create new materials through economic, direct reactor production of totally new polyolefins resins. Himont has commercialized this technology in its Catalloy polyolefin process. This proprietary process is designed to make maximum use of the reactor granule technology

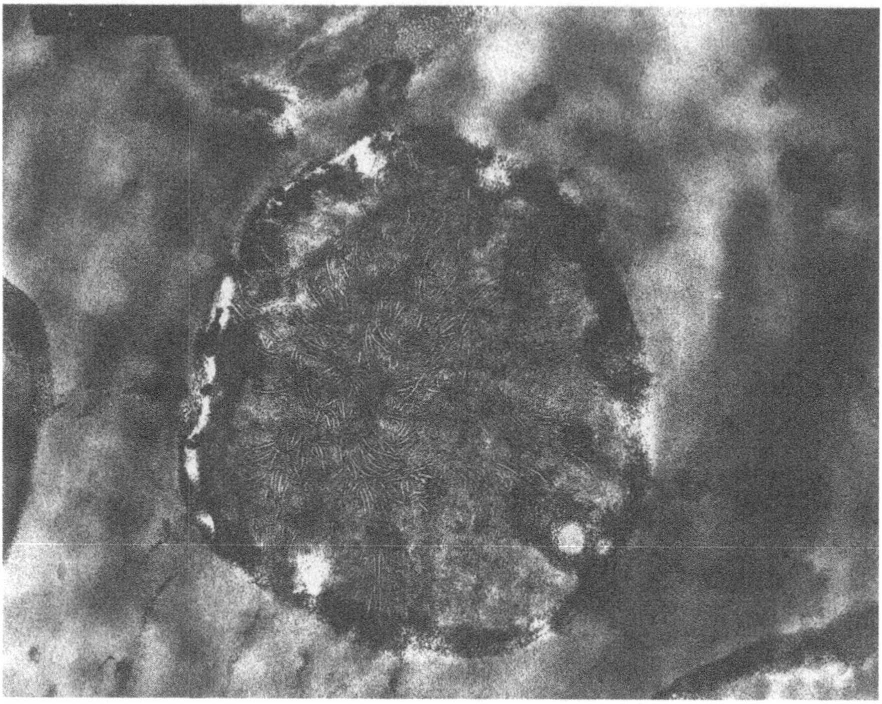

Figure 1.12 Cross-section of polypropylene–ethylene–propylene rubber–polyethylene heterophasic resin (× 30 000).

by allowing the repeated introduction of different monomers during the polymerization, generating multiphase, multipolymer alloys directly in the reactor. The resulting spherical copolymer resin is suitable for subsequent processing and fabrication with an extremely wide range of potential properties not possible from any other polypropylene process. The growing polymer particle has become the polymerization reaction medium, and it represents a real revolution in the development of Ziegler–Natta polymerization technology for the production of polypropylene copolymers. It removes almost all of the previous process constraints, allowing unique reactor-made copolymer resins to be produced with properties no longer limited by mechanical considerations of the process. The Catalloy process design was chosen on the basis of economics, mechanical reliability and reaction control criteria in order to maximize reactivity and productivity while at the same time producing the widest range of multipolymer structures.

Polypropylene random copolymers constitute an important class of plastic resin, particularly suitable for the packaging industry in films and containers requiring a balance of good clarity, flexibility and mechanical strength [57–59]. Recent market trends are towards film products with increasingly lower heat seal initiation temperatures, allowing faster, more economic processing. This property is related to the melting temperature of the polymer, which decreases with the insertion of the comonomer along the polymer chain.

The introduction of molecules of a second comonomer along with polypropylene chain promotes the formation of shorter, stereoregular blocks, allowing for the development of the crystalline γ-form, which melts at a lower temperature. Ethylene has been the normal comonomer for polypropylene random copolymers and results in a linear depression of melting point of approximately 5 °C per wt% of ethylene. The use of butene-1 as comonomer has been claimed to give unique properties to films that are of major commercial interest in the packaging industry [57]. A lower solubles content and a higher stiffness are claimed for any given melting point compared to propylene copolymers and ethylene. Both lower extractables content and higher stiffness are translatable to better processing characteristics. Higher stiffness relates to improved handling characteristics and the possibility of down gauging during film conversion. Lower extractables content is important in meeting US Federal Drug Administration (FDA) regulations for food packaging, which is often difficult for many low-melting ethylene-containing copolymers.

Copolymers of propylene with other α-olefins such as hexene-1 have been claimed to yield unique crystallization kinetics and a bulk morphology different from that of butene-1 and ethylene copolymers. Of significant interest is the change in crystallization temperature as a function of cooling rate. Regardless of crystallization temperature, hexene-1 copolymers appear to supercool to a significantly higher degree than ethylene copolymers, at rapid cooling rates. This can result in a higher degree of clarity in this film, which is another property of interest to the packaging industry.

The control of stereoregularity and the ability to contain low-melting, sticky resins within a hard polypropylene matrix allow the production of heterophasic polyolefinic alloys to be tailored to special end uses, such as additives for bitumen to improve processability and brittleness temperature. It requires a significant understanding and control of polymerization, polymer structure and compatibility to produce a polypropylene-based alloy which will give the required continuous phase necessary to satisfy the performance requirements of the application. [54, 60]. Figure 1.13 illustrates two polypropylene copolymer–bitumen blends in which the chemical structure of the polypropylene copolymer has been optimized to give either a bitumen continuous phase or a polypropylene continuous phase.

In applications where toughness combined with strength is an important criterion, such as in automotive bumpers, heterophasic copolymer blends of polypropylene and ethylene–propylene rubber were developed to compete with the polyester–polycarbonate blends that had previously displaced chrome-plated metal [61]. As mentioned earlier, such heterophasic

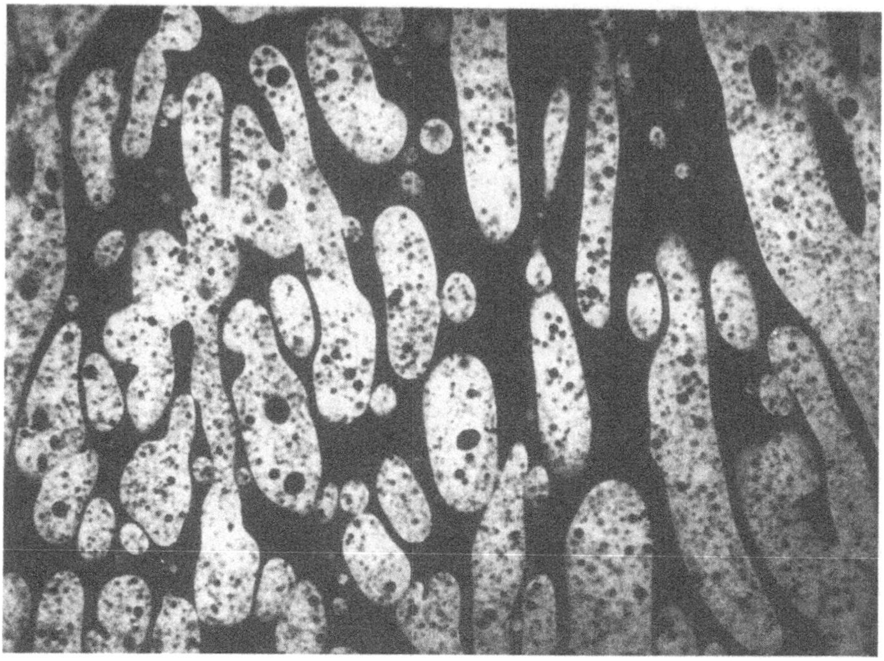

(a)

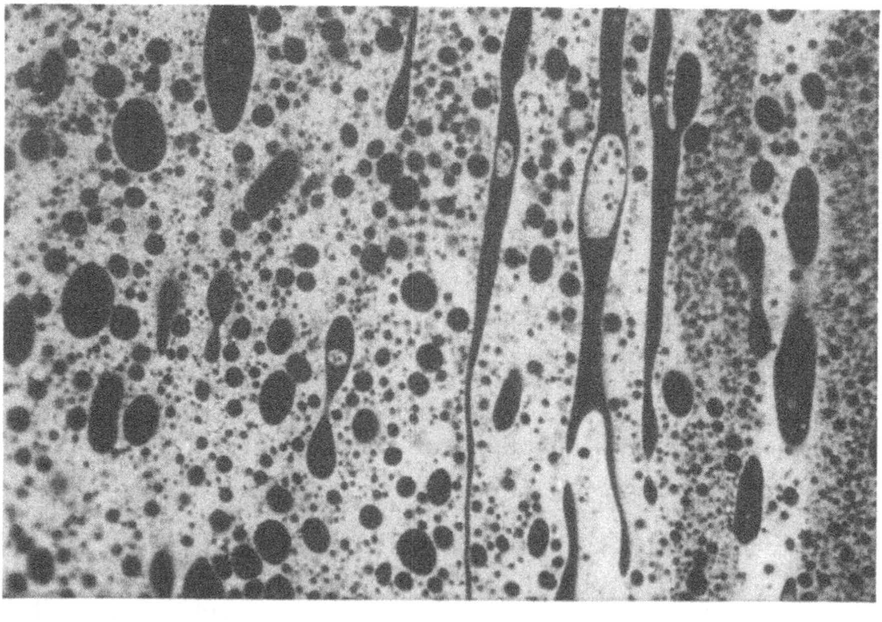

(b)

Figure 1.13 Polypropylene copolymer bitumen blends. Continuous phase (a) bitumen and (b) polypropylene (× 70).

polypropylene copolymer structures can show significant problems with stress whitening on impact of a moulded part. This whitening phenomenon, due to cavitational effects, is minimized by the addition of polyethylene to the heterophasic blend. The role of the polyethylene is to reduce residual stresses within the heterophasic rubber. During the melt cooling process, the volumetric contraction of the polypropylene matrix, due to crystallization, results in a compression of the heterophase made up of rubber and molten polyethylene. The subsequent polyethylene crystallization with consequent volumetric contraction promotes a progressive decrease in the heterophasic stress, eliminating the objectional stress whitening occurring on impact of the moulded part [60, 62–65].

The recent synthesis of polypropylene copolymers and alloys directly in the reactor rather than by extrusion melt blending has given resins with unexpected new properties, leading to new applications. These properties can be controlled by the chemical structure and physical properties of the second or third phase of the heterophasic copolymer or the matrix polymer itself. For instance, using the Catalloy process technology, an investigative program has been undertaken by Himont to synthesize a wide range of polypropylene alloys and blends in the reactor, which could significantly expand the property envelope of polyolefins and enable than to compete in applications with other, non-olefinic plastic resins. Novel, tough, flexible, impact resistant, clear polypropylene-based resins were synthesized to compete with polyvinyl chloride (PVC) in the packaging film industry. Resins were produced that have similar flexibility to PVC without the need for a plasticizer and have better clarity in film form. A surprisingly significant improvement in tear strength of these films compared to PVC was also noticed. Table 1.1 compares the properties of such polypropylene copolymers with those of PVC. The improvement in tear strength was found to be releated to the chemical structure and properties of the second phase in the heterophasic polypropylene copolymer, and the ability to produce and intimately disperse the polymer chains in the matrix polymer by the reactor granule technology. Films from such resins are finding increasing acceptance as geomembranes in pond liners and the lining and capping of landfills. The combination of puncture resistance and elevated temperature properties of such polypropylene copolymer films have also found acceptance as biohazard packaging bags that are required to be steam sterilized.

In summary, therefore, the development and commercial exploitation of polypropylene random and heterophasic copolymers can be claimed to have fuelled much of the dramatic growth in demand for polypropylene during the last decade. This expansion of the property envelope of polypropylene has enabled it to compete successfully in a growing range of applications with many other, less environmentally friendly, plastic resins. The recent ability to synthesize a wider variety of random and heterophasic polypropylene copolymers directly in the reactor will make polypropylene copolymers

Table 1.1 Properties of two polypropylene heterophasic copolymers compared with those of PVC

Property		*ASTM test number*	*PVC*	*Heterophasic polypropylene copolymers*	
Thickness (mm)		D-1593	0.76	0.76	0.76
Specific gravity		D-792	1.23	0.89	0.89
Tensile yield (MPa)	MD	D-882	dny[a]	10.1	6.6
	CD		dny	7.5	5.5
Elongation yield (%)	MD	D-882	dny	31	55
	CD		dny	20	41
Tensile break (MPa)	MD	D-882	15.9	11.4	17.9
	CD		16.5	7.9	20.9
Elongation break (%)	MD	D-882	430	631	800
	CD		405	560	880
Elastic modulus (MPa)	MD	D-882	13.4	308	102
	CD		14.1	171	65
Hydrostatic resistance (MPa)		D-751	0.56	0.76	0.55

[a] dny – did not yield.

increasingly the cost-effective resins of choice in the global plastic applications market place.

1.8 APPENDIX: SYMBOLS AND ABBREVIATIONS

ASTM	American Society for Testing and Materials
DSC	Differential scanning calorimetry
E	Ethylene
FDA	Federal Drug Administration (US)
NMR	Nuclear magnetic resonance
P	Propylene
PVC	Polyvinyl chloride

1.9 REFERENCES

1. Lieberman, R. and Barbe, P. C. (1988) Propylene polymers, in *Encyclopedia of Polymer Science and Engineering*, 2nd edn, Vol. 13, pp. 464–531.
2. Natta, G., Pino, P. and Mazzanti, G. (1954) Italian Patent 535 712, 8 June.
3. Natta, G., Pino, P. and Mazzanti, G. (1954) It. priority, 8 June.
4. Natta, G., Corradini, P. and Allegra, G. (1961) *Journal of Polymer Science*, **51**, 399–410.
5. Solvay (1972) French Patent, 2 130 231, 5 October.
6. Cossee, P. (1962) *Transactions of the Faraday Society*, **58**, 1226.
7. Arlman, E. J. and Cossee, P. (1964) *Journal of Catalysis*, **3**, 99–104.

8. Boor, J. (1979) *Ziegler–Natta Catalysis and Polymerization*, Academic Press, New York.
9. Rodriguez, L. A. M., Van Looy, H. M. and Gabant, J. A. (1966) *Journal of Polymer Science, Part A*, **4**, 1905–16.
10. Montedison (1978) US Patent 4 107 413, 15 August.
11. Montedison (1978) US Patent 4 107 414, 15 August.
12. Montedison (1978) US Patent 4 115 319, 19 September.
13. Guidetti, G., Zannetti, R., Ajo, D. *et al.* (1980) *European Polymer Journal*, **16**, 1007–15.
14. Simonazzi, T., Cecchin, G. and Mayr, A. (1987) Copolimeri del Polipropilene: da Miscele Polimeriche a Leghe per Sintesi, in Proc. VIII Convegno Italiano di Scienza delle Macromolecole (AIM), Milan, October.
15. Galli, P., Simonazzi, T. and Del Duca, D. (1988) *Acta Polymerica*, **39**, 81–90.
16. Galli, P., Cecchin, G. and Simonazzi, T. (1988) Advanced polymer properties: the new frontiers of the ideal Ziegler–Natta catalysts, in Proc. Frontiers in Macromolecular Science, IUPAC 32nd International Symposium, Kyoto.
17. Galli, P., Luciani, L. and Cecchin, G. (1981) *Angewandte Makromolekulare Chemie*, **94**, 63–89.
18. Begley, J. W. (1966) *Journal of Polymer Science*, **A1** 319–36.
19. Lawrence, R. L. and Chiovetta, M. G. (1987) Heat and mass transfer during olefin polymerization from gas phase, in Polym. React. Eng., 73, Hanser, Munich.
20. Ferrero, M. A. and Chiovetta, M. G. (1987) *Polymer Engineering and Science*, **27**, 1436–77.
21. Ferrero, M. A. and Chiovetta, M. G. (1987) *Polymer Engineering and Science*, **27**, 1448–60.
22. Ferrero, M. A. and Chiovetta, M. G. (1991) *Polymer Engineering and Science*, **31**, 886–903.
23. Ferrero, M. A. and Chiovetta, M. G. (1991) *Polymer Engineering and Science*, **3**, 904–11.
24. Kakugo, M., Sadatoshi, H. and Sakai, J. (1989) Morphology of nascent polypropylene produced by $MgCl_2$ supported Ti catalysts, in Proc. International Symposium on Olefin Polymerization Catalysts, Tokyo, pp. 345–54.
25. Gardner, G., Parsons, J. W. and Haward, R. N. (1978) *Journal of Polymer Science, Polymer Chemistry Edition*, **16**, 1683–96.
26. Spitz, R., Lacombe, J. L. and Guyot, A. (1984) *Journal of Polymer Science, Polymer Chemistry Edition*, **22**, 2625–40.
27. Keii, T. (1982) *Makromolekulare Chemie*, **183**, 2285–304.
28. Doi, Y., Murata, M., Yano, K. and Keii, T. (1982) *Industrial and Engineering Chemistry, Product Research and Development*, **21**, 580–5.
29. Cecchin, G. and Cometto, C. (1986) Catalyst developments for heterophasic copolymers, in Proc. Eurochem. Conference, Lahti, Finland, August.
30. Galli, P. and Haylock, J. C. (1991) *Progress in Polymer Science*, **16**, 443–62.
31. Barbe, P. C., Cecchin, G. and Noristi, L. (1987) *Advances in Polymer Science*, **81**, 1–81.
32. Cecchin, G. and Simonazzi, T. (1989) Polymeric alloys from functional polypropylene, in Proceeding Macromolecules 89, Oxford, September.
33. Himont Incorporated (1986) US Patent 4 602 077, 22 July.
34. Simonazzi, T., Cecchin, G. and Mazzullo, S. (1991) *Progress in Polymer Science*, **16**, 303–29.
35. Johnson, E. and Chowdhury, J. (1990) *Chemical Engineering*, May, 30–7.
36. Di Drusco, R. and Rinaldi, R. (1984) *Hydrocarbon Processing*, 113–7.
37. Rinaldi, R. *Chimica ed Industria* (Italy), **234**, 234–40.

38. Ficker, H. K. Goeke, G. L. and Powers, G. W. (1987) *Plastics Engineering*, February, 29–32.
39. Anonymous (1985) *European Chemical News*, 16 December, 19–20.
40. Seiler, E. (1989) Structure and properties of new polypropylene copolymers and reactor blends obtained by a special gas phase process, in Proc. Polypropylene: The Way Ahead, Madrid.
41. Avella, M. Martuscelli, E., Simonazzi, T. *et al.* (1986) *Makromolekulare Chemie*, **187**, 1927–43.
42. Cavallo, P., Martuscelli, E. and Pracella, M. (1977) *Polymer*, **18**, 42–8.
43. Martuscelli, E., Pracella, M. and Crispino, L. (1983) *Polymer*, **24**, 693–9.
44. Turner-Jones, A. (1971) *Polymer*, **12**, 487–508.
45. Guidetti, G., Busi, P., Giulianelli, I. and Zanetti, R. (1983) *European Polymer Journal*, **19**, 757–9.
46. Liberman, S., Ojeda, J., Franzese, R. and Fantazzini, C. (1991) Composition properties relationships in propylene–ethylene random copolymers, in Proceedings I Congress Brasilero de Polymeros, San Paulo.
47. Galli, P., Simonazzi, T. and Barbe, P. C. (1983) Ethylene–propylene copolymerization – Ziegler–Natta catalysis shows its virtuosity, in Proc. VI Convegno Italiano di Scienza delle Macromolecole (AIM), Pisa, October.
48. Himont Incorporated, US Patent 4 607 077, 19 August.
49. Takahashi, T., Mizuno, H., Thomas, E. L. *et al.* (1983) *Journal of Macromolecular Science Physics*, **B22** (3), 425–36.
50. Karger-Kocsis, J. and Kiss, L. (1987) *Polymer Engineering and Science*, **27**, 254–62.
51. Pukánszky, B., Tüdős, F., Kalló, A. and Bodor, G. (1989) *Polymer*, **30**, 1399–1406.
52. Pukánszky, B., Tüdős, F. Kalló, A. and Bodor, G. (1989) *Polymer*, **30**, 1407–13.
53. Varga, J. (1992) *Journal of Materials Science*, **27**, 2557–9.
54. Bucknall, C. B. (1977) *Toughened Plastics*, Applied Science, London.
55. Galli, P., Danesi, S. and Simonazzi, T. (1984) *Polymer Engineering and Science*, **24**, 544–54.
56. D'Orazio, L., Greco, R., Mancarella, C. *et al.* (1982) *Polymer Engineering and Science*, **22**, 536–41.
57. Ficker, H. K. and Walker, D. A. (1990) *Plastics and Rubber Processing and Applications* **14**, 103–8.
58. Kiang, W. (1985) *SPE Antec Preprints*, April, 467–9.
59. Galli, P., Milani, F. and Simonazzi, T. (1985) *Polymer Journal*, **17** (1), 37–55.
60. Simonazzi, T., Savadori, A. and Marchetti, E. (1982) Influence of structural and morphological parameters on impact strength of polypropylene block copolymers, in Proceedings IUPAC International Symposium, Athens.
61. Danesi, S. and Garagnani, E. (1983) Olefinic thermoplastic elastomers, in Proceedings International Rubber Conference, Wiesbaden.
62. Prentice, P. and Williams, J. G. (1982) *Plastics and Rubber Processing and Applications*, **2**, 27–32.
63. Fernando, P. L. and Williams, J. G. (1981) *Polymer Engineering and Science* **21**, 1003–9.
64. Karger-Kocsis, J. and Kuleznev, V. N. (1982) *Polymer*, **23**, 699–705.
65. Karger-Kocsis, J., Kalló A., Szafner, A. *et al.* (1979) *Polymer*, **20**, 37–43.

2

Primary spherulite nucleation in polypropylene-based blends and copolymers

Z. Bartczak, E. Martuscelli and A. Galeski

2.1 INTRODUCTION

The properties of a polymer can be extensively modified by physical mixing with another polymer. The properties of the resulting polymer blend depend not only on the composition and processing conditions but also on the physical state of each component at the temperature of application of the blend. Most of the theoretical and experimental investigations of the properties of polymer blends concern mainly systems containing amorphous components [1]. Recently, for both fundamental and practical reasons, blends with crystallizable components have received increasing attention [2].

Below the melting points of components, blends with crystallizable polymers constitute heterogeneous systems since the components separate from each other during crystallization. However, miscibility within the remaining amorphous phases of the components may still be possible. The phase separation induced by crystallization of components affects the mechanical properties of the blends, frequently causing their deterioration. On the other hand, the mechanical properties of blends are also determined by the properties of crystalline phase(s), including the overall crystallinity, crystalline morphology and the sizes of crystallites and their aggregates such as spherulites. Similarly to plain polymers, the mean size of spherulites in the blends is a very important factor influencing their mechanical properties, e.g. the yield stress and ultimate strength of the material [3]. Spherulite size

Polypropylene: Structure, blends and composites. Edited by J. Karger-Kocsis.
Published in 1995 by Chapman & Hall, London. ISBN 0 412 58430 1

is controlled mainly by the process of primary nucleation. The primary nucleation behavior depends on both the material properties, and the pressure and thermal treatment, i.e. thermodynamic conditions of melt processing and crystallization. In the case of blends the primary nucleation also depends strongly on the blend composition. Primary nucleation behavior in blends has been widely studied recently. This chapter summarizes the results of investigations of primary nucleation of spherulites in polymer blends containing crystallizing components, with particular attention focussed on the isotactic polypropylene based blends.

2.2 PRIMARY NUCLEATION IN POLYMERS

Crystals in polymers are grown from nuclei rather than formed uniformly in the entire volume of the material. In the primary nucleation phenomenon in polymers, three paths for nucleation can be distinguished: (i) homogeneous nucleation, which takes place if no preformed nuclei or foreign surfaces are present; (ii) heterogeneous nucleations, in which the nuclei form on foreign surfaces, which often reduces the nucleus size needed for stable crystal growth and thus enhance the nucleation process; and (iii) 'self-seeding' nucleation, unique to polymers, in which nucleation is caused by small polymer crystals that survived the melting or dissolution of the polymer sample [4].

More information about primary nucleation phenomena in isotactic polypropylene crystallized from the melt is given in Chapter 4 in Volume 1.

2.2.1 Observations of primary nucleation in polymer blends

Blends of miscible polymers

Microscopic observations of crystallization in blends of polymers that are miscible in the molten state have shown that the average size of spherulites in blends is generally larger than the size of spherulites in a plain polymer crystallized under the same conditions. Such observations, rather qualitative and frequently performed for samples crystallized nonisothermally with poorly controlled cooling rate, were reported for the following blends: isotactic polystyrene with atactic polystyrene (iPS/aPS) [5–7], iPS with poly (phenyleneoxide) (iPS/PPO) [8], polycaprolactone with poly(vinylchloride) (PCL/PVC) [7, 9], poly(ethyleneterephtalate) with poly(butyleneterephthalate) (PET/PBT) [7], poly(ethyleneoxide) with poly(methylenemethacrylate) (PEO/PMMA) [10] and high density polyethylene with linear low density polyethylene (HDPE/LLDPE) [11]. Observed increase of the average spherulite size in blends as compared to a plain polymer indicates that the number of primary nuclei of spherulitic crystallization in blends (calculated per unit volume of the sample) is lower than in plain crystallizing component. This behavior was interpreted as a result of the decrease of the concentration of crystallizable polymer and/or the lowering of supercooling resulted from a decrease of the melting temperature of crystal phase in a blend [9].

Quantitative investigations of primary nucleation phenomena under isothermal crystallization conditions, including the study of separate modes of the nucleation, were performed for two blends of miscible polymers: isotactic polypropylene with atactic polypropylene (iPP/aPP) [12, 13] and PEO/PMMA [14]. The samples of the iPP/aPP blend with various compositions were crystallized isothermally in the temperature range 122–131 °C [12]. Due to the high temperature of melt-annealing prior to crystallization (230 °C) all remaining traces of crystallinity were destroyed in the samples, so during the crystallization process self-seeding nucleation was excluded and only the heterogeneous nuclei were active (homogeneous nucleation for iPP requires larger undercooling and can be ignored at the applied crystallization temperature). Figure 2.1 shows the dependencies of nucleation density, *D* (the number of primary nuclei calculated per unit volume of a crystallizing polymer in the blend) on the blend composition, determined on the basis of Avrami analysis of differential scanning calorimetric (DSC) data. The results show that heterogeneous primary nucleation of iPP spherulites is strongly depressed when aPP is present in the blend. The stronger depression of heterogeneous nucleation at lower crystallization

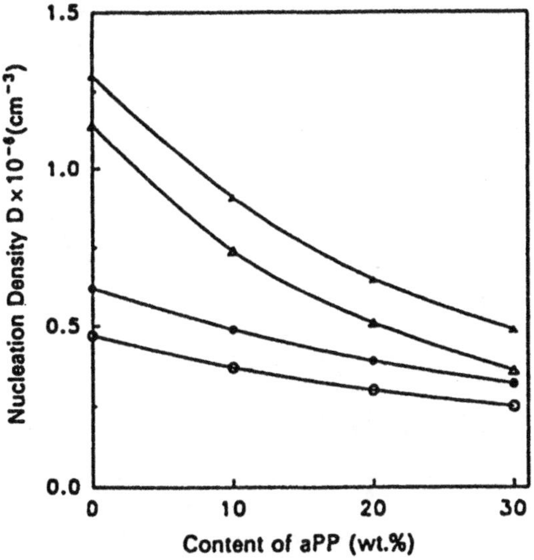

Figure 2.1 Dependence of nucleation density on the aPP content in iPP/aPP blends. Isothermal crystallization temperatures: (▲) 122 °C, (△) 125 °C, (●) 127 °C, (○) 131 °C. Temperature of melt-annealing: 220 °C. After [12].

temperatures suggests that in the crystallization of the blend only those heterogeneous nuclei are active that had to overcome the lowest energy barrier. Similar results were obtained in the studies of heterogeneous and self-seeding primary nucleation modes in a PEO/PMMA blend [14].

The primary nucleation of iPP crystallization in an iPP/aPP blend was also studied for isothermal crystallizations at very high undercooling [13]. In this study samples of plain iPP and iPP/aPP blends in the form of thin films were crystallized isothermally in a special cell at a temperatures ranging from 60 °C to 82 °C. Such extreme conditions (undercooling greater than 100 °C) were applied in order to investigate the homogeneous mode of primary nucleation in addition to heterogeneous nucleation. After completion of crystallization the average spherulite radius was determined in the samples using the small angle light scattering (SALS) technique (the average based on the 5th order moment). Figure 2.2 presents the determined values of the average spherulite radius in the blends as a function of crystallization temperature [13]. At higher crystallization temperatures (78–82 °C) heterogeneous nucleation is the dominant mode. The density of this nucleation is constant and independent of the temperature at the very high undercooling because of the saturation effect – all heterogeneities present in the sample volume induce the formation of heterogeneous nuclei. The homogeneous nucleation, although already active, still has too low a rate to influence the average size of spherulites in this temperature range. Such nucleation behavior results in the formation of the plateau on the spherulite radius versus crystallization temperature curve at temperatures higher than about 78 °C. However, with the decrease of the crystallization temperature the rate of homogeneous nucleation increases rapidly, whereas the density of

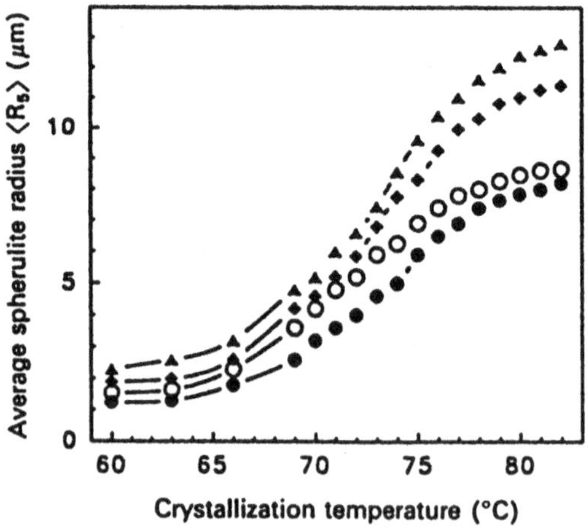

Figure 2.2 Dependence of the average spherulite radius $\langle R_5 \rangle$ on the crystallization temperature in samples of iPP and iPP/aPP blends: (•) iPP, (○) iPP/aPP 9:1, (♦) iPP/aPP 8:2, (▲) iPP/aPP 7:3. After [13].

heterogeneous nucleation remains constant, so finally the homogeneous nucleation acquires control of the average spherulite size in the sample. The further rapid increase of the homogeneous nucleation rate with the decrease of crystallization temperature induces a significant decrease in the average spherulite radius, clearly seen in Figure 2.2. In [13] a description of the homogeneous nucleation mode in blends was proposed. The rates of homogeneous nucleation in plain iPP and iPP/aPP blends were estimated based on the data presented in Figure 2.2 [13]. The results of these estimations are presented in Figure 2.3. It can be seen that for all crystallization temperatures applied, the rate of homogeneous nucleation in iPP/aPP blends decreases with increasing aPP concentration in the blend. The changes of the nucleation rate are much stronger than the changes of concentration of iPP in the blend. This shows that the decrease in the number of primary nuclei in miscible blends of iPP with aPP is not only the result of simple dilution of the crystallizing component by its amorphous counterpart, but that there are also other factors influencing primary nucleation in miscible systems. The problem was discussed in [13] and will be considered briefly later in this chapter.

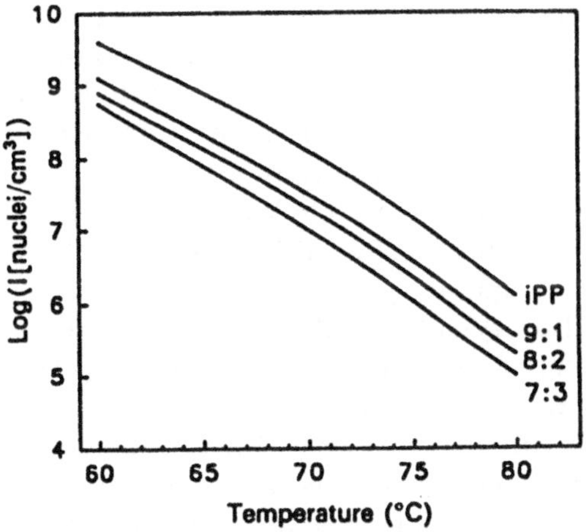

Figure 2.3 Rates of homogeneous nucleation in iPP and iPP/aPP blends, calculated per unit volume of iPP in the blend, plotted against crystallization temperature. After [13].

Blends showing phase separation (immiscible or partially miscible components)

The primary nucleation of spherulites in blends of immiscible or partially miscible polymers was observed and studied mainly in blends containing iPP as a major, crystallizing component (only in a few studies were both blend components able to crystallize) [15–38]. Only a few papers report the observations of primary nucleation in blends of other crystallizable polymers: polyamide-6 (PA-6) with ethylene–propylene rubber (EPM) functionalized by grafting on its backbone maleic anhydride (PA-6/EPM-g-MA) [39], isotactic polybutene-1 with low density polyethylene (iPB/LDPE) [40], poly(phenylenesulfide) with HDPE (PPS/HDPE) [41, 42], with PET (PPS/PET) [43] and with poly(tetrafluoroethylene) (PPS/PTFE) [42], as well as a blend of poly(oxymethylene) (POM) with urethane elastomer [44]. In the referenced studies crystallizing polymers formed the continuous phase (matrix) whereas the second component was dispersed in the matrix in the form of droplet inclusions.

Even at a first glance the primary nucleation in blends with phase separated components seems to be more complicated than that in miscible systems. Depending on the blend composition and/or experimental conditions various observations have been reported. Both increases and decreases of the nucleation density in blends, as compared to plain crystallizing polymers, have been reported. The results published on primary nucleation are summarized below.

The iPP/iPB blend was regarded as a blend of miscible polymers. More detailed investigations [15] have shown however, that this blend is phase separated and its components are only partially miscible. In the iPP/iPB blend both components are crystallizable, but iPP crystallizes first, at a higher temperature. Siegmann [20] studied the morphology of the iPP/iPB blend crystallized nonisothermally. He found that the size of spherulites in blends is generally lower than that in both plain iPP and plain iPB crystallized under similar conditions, except for the case of 50:50 composition. In this case a nonspherulitic morphology was revealed. Detailed study of isothermal crystallization of iPP in iPP/iPB blends (at a temperature of crystallization high enough to prevent crystallization of iPB) showed, however, that iPP also crystallizes spherulitically in the case of a 50:50 blend [15]. It was found that the density of heterogeneous nucleation (i.e. the number of heterogeneous nuclei per unit volume of iPP in the blend) decreases only slightly with the decrease of concentration of iPP in the blend (Figure 2.4). Similarly to miscible blends, discussed in the previous section, in blends of partially miscible polymers, nucleation at lower crystallization temperature is more strongly depressed than at higher temperatures. This suggests that during crystallization in a blend only the most active potential heterogeneous nuclei retain their activity.

Completely different nucleation behavior was found in other partially miscible polymer blends, namely blends of iPP with various elastomers: ethylene–propylene rubber (iPP/EPM) [17, 28–32], ethylene–propylene diene terpolymer (iPP/EPDM) [30, 31, 38] polyisobutylene (iPP/PIB) [29, 30] and *trans*-polyoctylene (iPP/TOR) [36]. In these investigations, in which DSC, SALS and microscopic techniques were used, a dramatic increase in the number of nuclei in the blends with increasing elastomer concentration in the system was found. The exemplary results obtained on

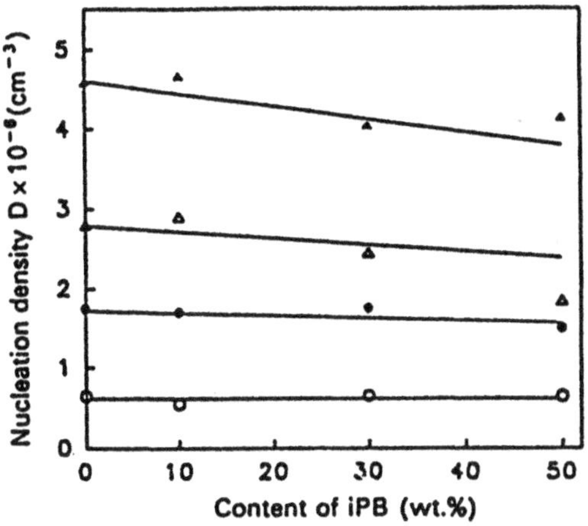

Figure 2.4 Dependence of the nucleation density in the iPP/iPB blends on the blend composition. Crystallization temperatures (▲) 119 °C, (△) 123 °C, (●) 125 °C, (○) 130 °C. Temperature of melt-annealing: 220 °C. After [15].

the basis of microscopic observations of isothermally crystallized blends of iPP with various types of EPM [29, 30] and EPDM [30] are presented in Figure 2.5. The changes in the number of nuclei in blends, as compared to plain iPP, were initially attributed to some activity of the elastomer as a nucleating agent for iPP crystallization. Through detailed studies of primary nucleation behavior in the iPP/EPM blend [17], it was discovered that the nucleation behavior in that blend is more complicated than reported in previous studies [28–32]. Drastic changes in the nucleation process in the blend were found not only with the variation of the composition but also with the variation of both the mixing time and the thermal history of the samples prior to their crystallization. To obtain samples of the blend the components were mixed in the molten state in a mixing extruder. Two sets of samples with the same composition but with different duration of melt-mixing (samples mixed one and twice in the extruder) were prepared and examined for the course of primary nucleation in isothermal crystallization. Moreover, for each sample three different temperatures of melt-annealing prior to crystallization were applied: 190 °C, 230 °C and 250 °C. These temperatures were chosen in order to study: (i) the heterogeneous mode of primary nucleation at $T_a = 230$ °C, sufficient to eliminate self-seeding nucleation in iPP, (ii) the self-seeding mode at $T_a = 190$ °C; at this temperature of melt-annealing self-seeding nuclei are not destroyed, so during crystallization both heterogeneous and self-seeding nucleation modes are active, and (iii) the influence of overheating and partial degradation on nucleation behavior, at $T_a = 250$ °C. The results of the investigations are presented in Figure 2.6. The densities of nucleation (recalculated to the volume of iPP in the blend) are plotted as a function of blend composition for various mixing conditions, melt-annealing and crystallization

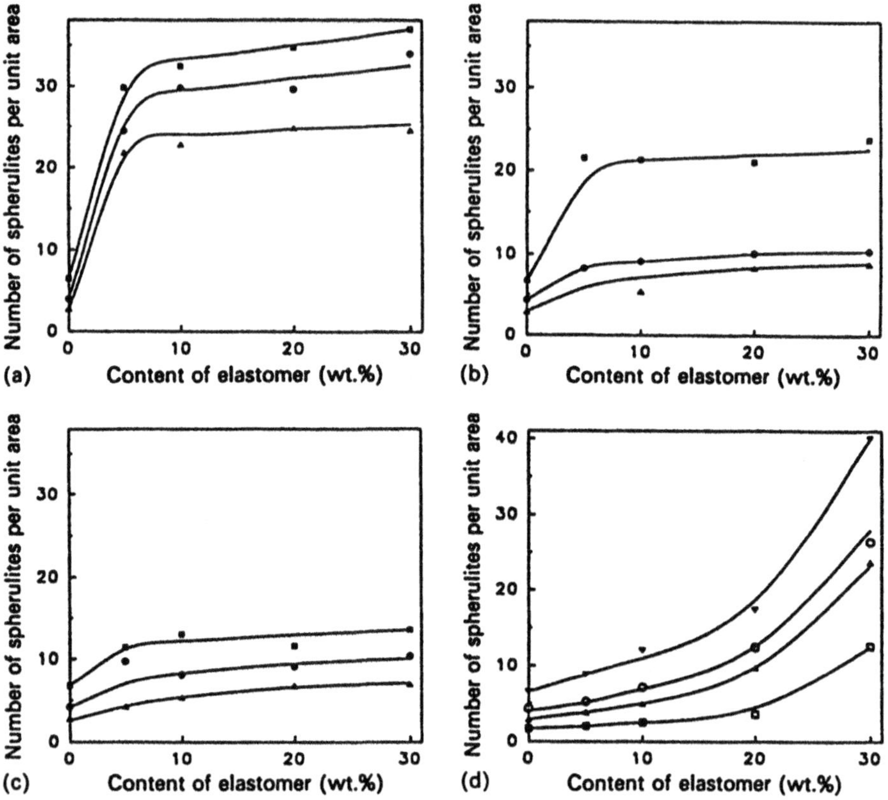

Figure 2.5 Number of spherulites per unit area as a function of elastomer content in thin film samples of iPP/elastomer blends with different types of elastomer crystallized at different crystallization temperatures: (■) 119 °C, (▼) 123 °C, (●) 125 °C, (○) 127 °C, (▲) 131 °C, (□) 135 °C. The grades of elastomer: (a) Epcar EPM, (b) Dutral EPM, (c) Buna EPM, (d) EPDM. After [29, 30].

temperatures. As can be seen, the nucleation density increases with increasing concentration of EPM in the blend for all conditions of sample treatment, except for the blends mixed once and annealed prior to isothermal crystallizations at 190 °C (Figure 2.6a). Additional experiments showed [17] that this unexpected behavior is caused by a significant depression of self-seeding nucleation mode in blends due to partial miscibility of components and an increase in the number of heterogeneous nuclei (compare Figure 2.6a and c)

It was demonstrated that the increase of nucleation density with increasing EPM content in the blend and with increasing mixing time observed for other conditions (see Figure 2.6) is not caused by the nucleation activity of EPM on the iPP crystallization, but is a result of increased number of heterogeneous nuclei. The impurities, which act as heterogeneous nuclei, migrate from the EPM phase into the iPP during preparation of the blend

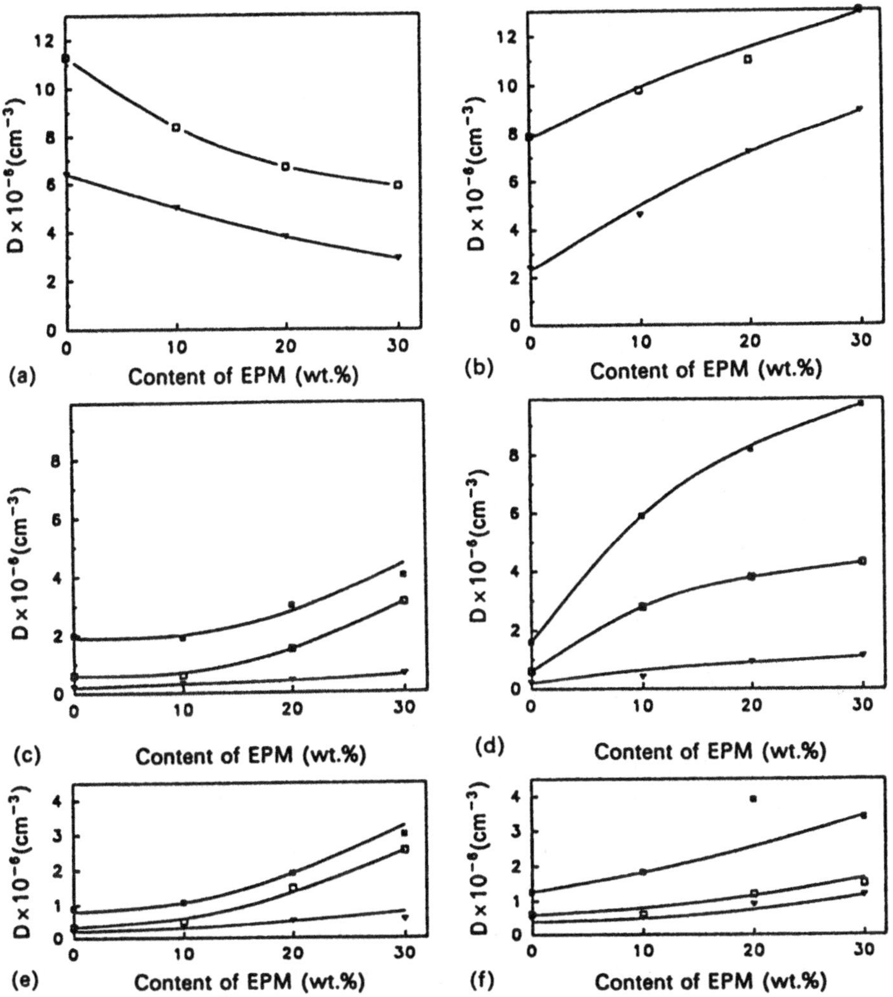

Figure 2.6 Dependence of the nucleation density in the iPP/EPM (Dutral) blends on the blend composition. Crystallization temperatures: (■) 119 °C, (□) 125 °C, (▾) 135 °C. Samples mixed (a, c, e) once and (b, d, f) twice in the extruder. Temperature of melt-annealing: (a, b) 190 °C; (c, d) 220 °C; (e, f) 250 °C. After [17].

by melt-mixing [17]. This phenomenon is discussed in detail in Section 2.3.1. It was also shown that partial degradation caused by overheating of the blend induces the decrease of the nucleation density (heterogeneous) in plain iPP as well as in blends (compare Figure 2.6c and e, and d–f).

The decrease of spherulite size induced by the increase of nucleation density was also reported for blends of iPP with other elastomers, namely with styrene–butadiene rubber (iPP/SBR) [28, 31], with polyisoprene (iPP/IR) [28] and with TOR (iPP/TOR) [36], as well as in blends of poly(oxymethylene) (POM) with urethane elastomer [44] and of PA-6 with ethylene–propylene rubber functionalized by grafting on its backbone maleic anhydride (PA-6/EPM-g-MA) [39]. On the other hand, the study of isothermal crystallization of the blend of iPP with polybutadiene (iPP/BR) showed a significant decrease in the heterogeneous nucleation density in the blend compared to plain iPP [33].

Crystallization, including the primary nucleation phenomena, proceeding in blends of iPP with LDPE, HDPE and LLDPE was studied by several authors [16, 18, 21–27, 35, 37]. It was found for the blend of iPP with LDPE that the number of heterogeneous primary nuclei of iPP crystallization, active during isothermal crystallization at a temperature above the temperature range for crystallization of LDPE, decreases with increasing concentration of LDPE in the blend [16, 21]. The nucleation densities determined in [21] for the blends if iPP with LDPE are presented in Figure 2.7. On the other hand, Teh [22] reported a decrease of spherulite size (i.e. an increase of nuclei number) in iPP/LDPE blends compared to plain iPP in samples crystallized nonisothermally.

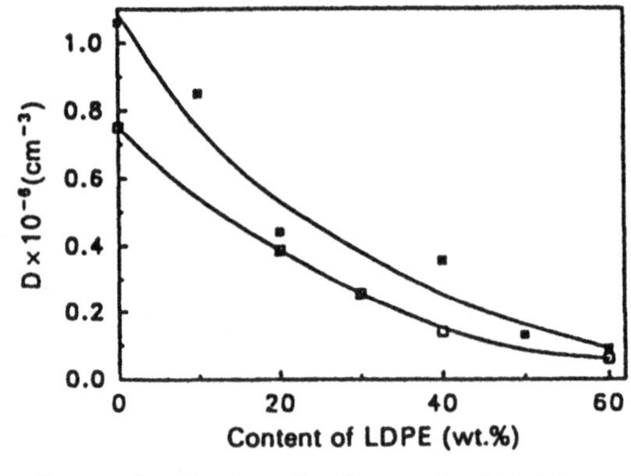

Figure 2.7 Dependence of nucleation density on the LDPE content in iPP/LDPE blends for isothermal crystallization temperatures: (■) 125.8 °C, (□) 128.8 °C. After [21].

Similar nucleation behavior was reported for iPP/LLDPE blends [37]: the number of primary nuclei of iPP spherulites decreased with increasing LLDPE content in blends crystallized under isothermal conditions.

Results concerning the primary nucleation in iPP/HDPE blends have been reported by several authors [18, 23–26]. Lovinger and Williams [23] observed a strong decrease of spherulite size (an increase of nucleation density) in samples of iPP/HDPE blend as compared to plain iPP or plain HDPE in samples crystallized under nonisothermal conditions after melting at 200 °C. Similar observations were reported by Noel and Carley [25].

Studies of primary nucleation behavior in iPP/HDPE blends under isothermal crystallization conditions [18] showed changes in heterogenous nucleation compared to plain iPP, whereas the self-seeding nucleation of the blends was almost unaffected. It was found that at temperatures higher than 127 °C, at which only the crystallization of iPP was significant and HDPE remained in a molten state, the density of heterogeneous nucleation of iPP spherulites decreased with increasing concentration of HDPE in the blend, similarly to iPP/LDPE [16] and iPP/LLDPE [37] blends. However, at lower temperatures, at which crystallization of both components can proceed at the same time, the density of heterogeneous nucleation of iPP spherulites strongly increases with increasing concentration of HDPE in the blend (in the range of concentrations studied, iPP constitutes the continuous matrix phase). The above described dependencies of nucleation density on the composition and the temperature are shown in Figure 2.8.

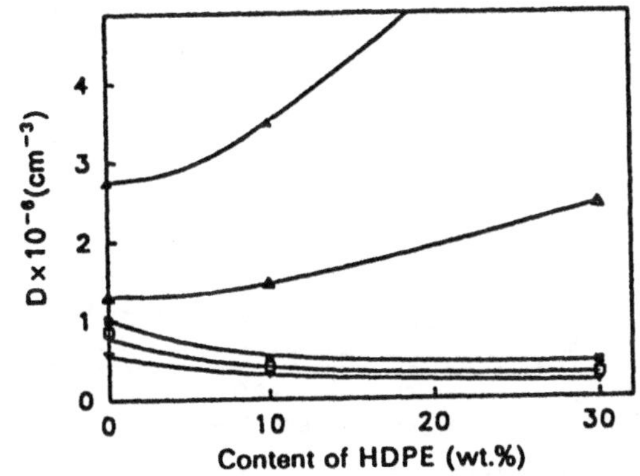

Figure 2.8 Dependence of nucleation density on the HDPE content in iPP/HDPE blends for isothermal crystallization temperatures: (▲) 123 °C, (△) 127 °C, (■) 129 °C, (□) 131 °C, (▾) 135 °C. Temperature of melt-annealing: 220 °C. After [18].

Another pair of immisicible polymers in which primary nucleation was studied was the blend of iPP with aPS (iPP/aPS) [19]. In this investigation an increase in the number of heterogeneous nuclei of iPP crystallization with increasing content of aPS in the blend was found. The increase is more pronounced for blends mixed in the molten state for a longer period of time (Figure 2.9). The density of self-seeding nucleation in iPP/PS blend also showed an increase with increasing concentration of PS in the blend.

In the other group of blends of polymers with the separation of components the minor, dispersed component is able to crystallize, while the matrix is amorphous or crystallizes in another temperature range. Crystallization of the minor component in the blend was reported in several papers: blends of SBR with iPP [45], EPDM with iPP [38], aPS with PE [46, 47], PE with POM [48], poly(vinylidenefluoride) (PVDF) with PA-6 and PBT [49] and iPP with PEO [50]. It was found in these investigations that if the crystallizing component is finely dispersed in the matrix its crystallization

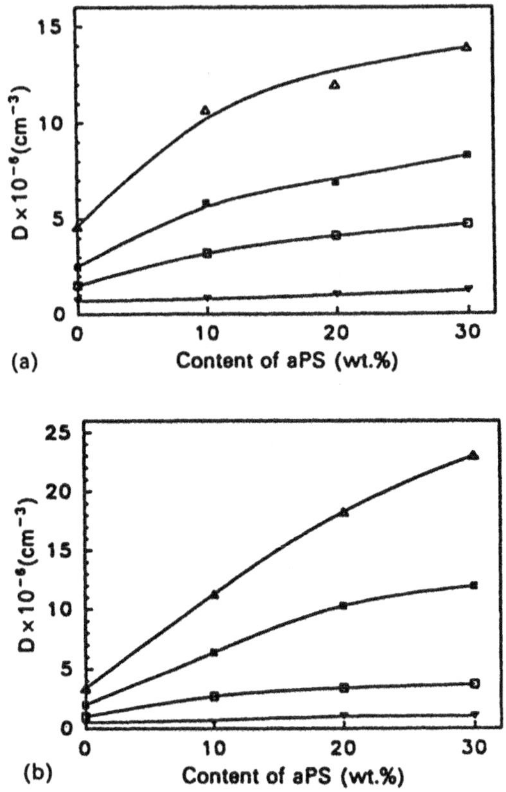

Figure 2.9 Dependence of nucleation density on the PS content in iPP/PS blends for samples mixed (a) twice and (b) three times in an extruder. Temperatures of crystallization: (△) 119 °C, (■) 123 °C, (□) 125 °C, (▾) 130 °C. Temperature of melt-annealing: 220 °C. After [19].

during cooling is often split into several distinct steps ('fractionated crystallization'). The origin of these steps is the primary nucleation in isolated molten inclusions by different nucleating species. The type of crystallization is similar to the crystallization in so-called 'droplet' experiments, in which polymer is finely dispersed in the form of small droplets in an inert medium [51, 52]. Similarly to droplet experiments, the steps of fractionated crystallization observed as separate DSC peaks can extend over a temperature range even greater than 60 °C [45]. The occurrence of this type of crystallization depends strongly on the state of dispersion of blend components. If the size of inclusions of crystallizing polymer is large, then the crystallization kinetics do not differ markedly from that observed for the bulk material. Splitting into separate steps occurs when the inclusions are sufficiently small, so that their number is comparable to the number of nucleating heterogeneities. In that situation all heterogeneities distributed in droplets get a chance to activate a heterogeneous nucleus, whereas in a bulk sample only the most active heterogeneities (i.e. those that induce the strongest decrease of the energy barrier for nucleation) give rise to spherulites. Weaker heterogeneities lose the competition and remain inhibited for kinetic reasons.

It is also believed that for sufficiently fine dispersion homogeneous nucleation of crystallization in some fraction of the inclusions may be observed [45, 50]. This could be possible when the number of crystallizing inclusions dispersed in the matrix is larger than the number of heterogeneities able to induce the nucleation. Detailed discussion of the phenomenon of fractionated crystallization, including estimation of the number of nuclei and energy considerations is given in [49].

2.2.2 Primary nucleation in propylene copolymers

There is very little experimental data in the literature concerning primary nucleation of copolymers containing propylene crystallized from the melt. Several authors found for isotactic propylene copolymerized with relatively small amounts of ethylene [53, 54] and butene-1 [55] that the random incorporation of comonomer units into iPP chains promotes crystallization of polypropylene in the γ-form instead of the usual α-form. Marigo *et al.* [54] found in studies of fractionated samples of propylene–ethylene random copolymer that increasing content of the ethylene comonomer promoted crystallization in the γ-form. They observed that slow-cooling crystallization of the copolymer fraction with the highest ethylene comonomer content (6 wt%) generated crystallites almost entirely in the γ-form. The phenomenon of the capability of the random copolymers to crystallize preferentially in γ-form is not yet completely understood. It was suggested [54] that the incorporation of the comonomer units into regular sequences of the iPP chain is responsible for an increase of the number of defects along the chain, which in turn may promote the crystallization of iPP in the γ-form.

Studies of the kinetics of crystallization of propylene copolymers demonstrated that the crystallization rate of random copolymers of propylene with ethylene [53] and butene-1 [56] is substantially reduced compared to that of iPP homopolymer. This is probably due to a decrease of both primary nucleation density and the growth rate of spherulites.

Investigations of crystallization of propylene blocks in ethylene–propylene diene block copolymers revealed that under nonisothermal conditions the temperature of the DSC peak respective to crystallization of propylene blocks in copolymer did not differ markedly from the temperature of the peak of iPP crystallization in the blend iPP/random EPDM copolymer, provided that the fraction of isotactic sequences of propylene in the blend was close to that in the block copolymer [38]. This result may suggest that the primary nucleation and the spherulite growth rate in block copolymer were not retarded in block copolymer compared to those in the blend, although the temperature of the crystallization peak under nonisothermal conditions is not a precise parameter for evaluation of the nucleation and growth parameters (it depends primarily on the experimental conditions, such as the cooling rate, conditions for heat transfer, etc.) and gives only a very qualitative description of the crystallization process. A more precise study of crystallization in block copolymers was performed by Drzewinski [33], who determined the densities of primary nucleation of iPP blocks in propylene–butadiene block copolymer under isothermal crystallization. On the basis of DSC data analysis he found that the density of primary nucleation in block copolymer is several times higher than that in iPP homopolymer crystallized under similar conditions.

2.3 DISCUSSION

2.3.1 Blends

The results surveyed in Section 2.2 indicate that the presence of other components in blends has a great effect on the primary nucleation of the crystallizing component. It was shown that practically all modes of primary nucleation – homogeneous, heterogeneous and self-seeding – are influenced in blends.

The influence of the second polymeric component in a blend on primary nucleation of the crystallizing component depends on: (i) the chemical structure of the second polymer, (ii) its physicochemical properties including the miscibility and its ability to crystallize and, of course, (iii) its concentration in a blend. In some blends of immiscible or partially miscible polymers the conditions of the mixing process appear to be an important factor from the point of view of primary nucleation.

In [13] the influence of miscibility of blend components on homogeneous nucleation was considered. The following equation, being the modification of the Turnbull–Fisher equation for nucleation rate in homopolymers, was proposed for the rate of homogeneous nucleation in blends of miscible polymers:

$$I = I_{ob}\exp\left(-\frac{U^*}{R(T-T_{\infty b})}\right)\exp\left(-\frac{32\sigma^2\sigma_e}{(\Delta g')^2 kT}\right) \tag{1}$$

where $I_{ob} = I_o c'$; I_0 is the nucleation constant representative for crystallizable polymer and c' is a factor dependent on the concentration of crystallizable polymer in the blend, c. The value of c' is within the limits $c \leqslant c' \leqslant 1$. If the nucleation rate is calculated per unit volume of crystallizable component the nucleation constant I_o should be used instead of I_{0b}. $U^* = 17.25$ kJ/mol is a constant characteristic for the activation energy of viscous flow. $T_{\infty b} = T_{gb} - 51.6$ K, where T_{gb} is the glass transition temperature of the blend. $\Delta g' = \Delta g_f + \Delta g_m$ is the free enthalpy of fusion in the mixture (Δg_f is the free enthalpy of melting of the crystal phase and Δg_m is the free enthalpy of mixing of blend components, both calculated per unit volume of crystallizing component in the blend). In Equation 1 the concentration of crystallizing polymer, c, does not appear explicitly. Instead, there is the c' parameter depending on c. The reason for this is that the nucleation events occur preferably in places where the local concentration of crystallizing segments c' fluctuates to values larger than the macroscopic concentration c. Support for this concept is given by the investigation of crystallization of linear polyethylene from solution, which has shown that in the case of secondary nucleation $c' \approx c^{1/3}$ [57]. One can expect that a similar relation should be observed for primary nucleation in multicomponent systems. The changes in the condition of the transport of crystallizing segments across the crystal–melt interface in the blend are taken into account in the proposed equation by considering the changes in glass transition temperature of the system (first exponential term). The free enthalpy $\Delta g'$ instead of Δg_f appears in the last term of the proposed equation. It takes into account the necessity of the local separation of blend components during the formation of a stable nucleus. The blend components are compatible in the molten state, thus $\Delta g_m < 0$. Since $\Delta g_f > 0$, $\Delta g' = (\Delta g_f + \Delta g_b) < \Delta g_f$, and therefore the energy barrier for the formation of a critical nucleus in the blend of compatible polymers is greater than in plain polymer. This results in a decrease of nucleation rate in the blend of compatible polymers as compared to a plain polymer. Equation 1 seems to be a good approximation of primary nucleation in the miscible blend of two polypropylene isomers: iPP and aPP [13]. However, it is probably not satisfactory in the case of a blend of two miscible but chemically different polymers, unless the variation of surface energies of the crystal σ and σ_e in a blend is considered. These energies are different for crystals growing in a blend than for those in plain homopolymer because the melt surrounding the crystal is a homogeneous mixture of two chemically different components. This effect was taken into consideration in the description of crystallization in blends proposed recently by Rostami [58].

The influence of miscibility of blend components on the heterogeneous mode of nucleation as well as on the self-seeding can also be considered. However, these nucleation modes are much less sensitive to the change in the condition of transport. Thus, the change in the energy barrier of the formation of critical nucleus caused by the decrease in the value of free enthalpy, $\Delta g'$, is practically the only important factor. The increase of that barrier, resulting from the separation of components surrounding the growing nucleus, causes some of the heterogeneities to become unable to induce instantaneous nucleation (although in plain polymer they could induce such nucleation) but a time is required for the formation of a stable nucleus on their surfaces. That time may be longer than the time necessary for the crystallization of the sample which begins on other more active nuclei. In this way such heterogeneities, being potential nuclei, are excluded from the crystallization process for kinetic reasons. As a result, the number of heterogeneous nuclei active in the blend is lower than in plain crystallizing component, and decreases with increasing concentration of a second polymer in the blend. Potential nuclei which in plain polymer are less active (higher energy barrier) should lose their activity in blends. This in turn should induce a larger depression of heterogeneous nucleation in blends at lower crystallization temperatures. Such a behaviour was indeed reported for iPP/aPP [12] and PEO/PMMA [14] miscible blends.

Similar trends, although less pronounced, were also observed in iPP/iPB blends [15], the components of which are only partially miscible. In those blends the discussed influence of miscibility on heterogeneous nucleation was probably limited to the volume of interfacial layers between components, whereas in the volume of iPP matrix, free of iPB molecules, heterogeneous nucleation was probably not disturbed. As a result the nucleation density in the sample of the blend as a whole was only slightly lowered with respect to that in plain iPP.

Based on similar considerations, one can also draw similar conclusions for the self-seeding mode of primary nucleation, both in the case of full miscibility of blend components and in the case of partial misciblity. The experimental evidence of such expected behavior of self-seeding comes from the results of investigations of nucleation in samples of iPP/EPM blend mixed once in the extruder [17]. The decrease in the number of self-seeding nuclei in the blend with increasing rubber concentration was in fact observed (see Figure 2.6a). In the samples of the blend mixed twice, this was not observed because of large changes in heterogeneous nucleation related to prolonged mixing screening the changes in self-seeding [17].

This concept of the influence of the second polymer component on nucleation behavior of crystallizing polymer takes into consideration only a miscibility of components. Thus, it is unable to describe changes of the primary nucleation behavior in cases of blends of immiscible polymers.

In order to explain the nucleation behavior in iPP/LDPE blend (see Figure 2.7) Galeski, Bartczak and Pracella formulated in [16] the hypothesis of migration of heterogeneities (particles of impurities and/or additives), present in both components before blending, from one polymer toward the other during the process of mixing in the molten state. The driving force for such migration across the interface would be the difference in interfacial free energies of the impurities with respect to both molten components. The tendency to minimize the interfacial energies would favor the migration of heterogeneities toward that blend component in which their interfacial energies are lower. In the case of iPP/LDPE blend the migration occurs from iPP towards LDPE. As a result, the iPP phase in blend becomes poorer in heterogeneities constituting the potential heterogeneous nuclei compared to plain iPP, so the density of heterogeneous nucleation of iPP spherulites in the blend is lower than in plain iPP. This was indeed observed experimentally.

The hypothesis of migration was confirmed in the specially designed crystallization experiment, in which the nucleating agent was introduced at the beginning to the volume of one of the blend components, then the samples were melt-mixed in repetitive cycles in a mixing device. After each mixing cycle a small amount of the blend was collected and the nucleation behavior in that sample was investigated. Figure 2.10 illustrates the results of the experiment, performed for iPP/LDPE 8:2 blend nucleated with sodium benzoate. It can be seen that the migration of the particles of nucleating agent from iPP to LDPE causes a dramatic decrease of nucleation density with increasing mixing time in the sample in which nucleating agent was initially introduced to iPP phase. The particles of sodium benzoate do not migrate in the opposite direction; the nucleation behavior of the sample in which the nucleating agent was introduced initially to LDPE is close to that observed in a control iPP/LDPE blend prepared without nucleating agent.

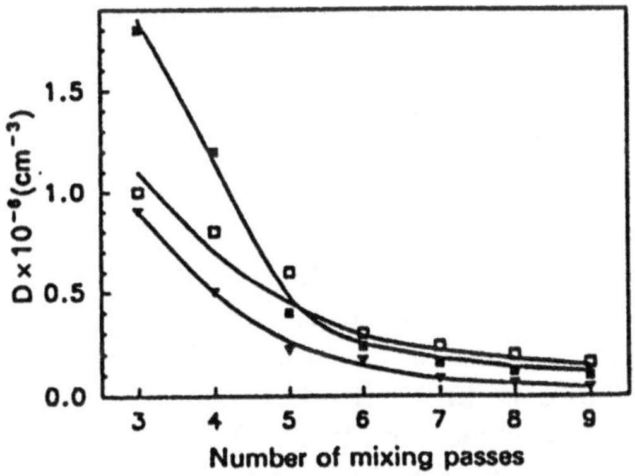

Figure 2.10 Plots of nucleation density versus mixing time for 8:2 iPP/LDPE blends containing 0.2 wt% sodium benzoate, crystallized isothermally at 132 °C: (■) iPP*/LDPE, (□) iPP/LDPE*, (▾) iPP/LDPE. The asterisk indicates the polymer to which the sodium benzoate was initially introduced. After [16].

The phenomenon of a migration of impurities during the mixing process was observed and confirmed by similar experiments using nucleating agents in other polymer blends. Migration in the same direction as in the iPP/LDPE blend was found in iPP/HDPE blend [18]. Migration in the opposite direction, i.e. towards polypropylene, was found in iPP/EPM [17] and iPP/aPS [19] blends. In the last two blends the migration induces significant increase of the density of heterogeneous nucleation in blends with increasing content of the second component and increasing mixing time.

The migration was also suggested as a possible cause of changes of primary nucleation density in blends of iPP with polybutadiene (PBd) [33], iPP with LLDPE [37], POM with urethane elastomer [44] and PVDF/PA-6 and PVDF/PBT blends [49].

Another factor influencing the heterogeneous nucleation in blends is the ability of a second blend component to crystallization. If the second polymer dispersed in a crystallizable matrix can crystallize under similar conditions as the matrix polymer, then the crystals of a second polymer grown at interfaces can act as nucleating agent for crystallization of the matrix. Such behavior was found in the iPP/HDPE blend [18]. HDPE dispersed in the iPP/HDPE blend crystallizes below 125–127 °C with a rate close to that of crystallization of iPP. Crystalline HDPE inclusions dispersed in iPP matrix induce additional nucleation of a number of iPP spherulites. As a result, the nucleation density of iPP spherulites in the iPP/HDPE blends increases with increasing content of HDPE in the blend during crystallization below 127 °C [18]. It is interesting to note that because of the phenomenon of migration the HDPE inclusions become richer in impurities, thus its crystallization is enhanced, which in turn induces a large increase of the nucleation density of iPP in that blend because of the large number of HDPE crystallized inclusions contacting iPP melt.

The influence of crystallization of a dispersed polymer on the matrix crystallization can probably explain the decrease of spherulite size in nonisothermally crystallized samples of the blends reported for iPP with LDPE [22], iPP with HDPE [23] and iPP with iPB [20]. During fast nonisothermal crystallization a simultaneous crystallization of both components is also possible. Already crystallized inclusions of a dispersed polymer accelerate the crystallization of a matrix acting as nucleating agent and induce the formation of additional spherulites. Finally, the average spherulite radius in those blends becomes smaller than in plain iPP crystallized under the same conditions. The influence of crystallization of a dispersed component on the rise of crystallization temperature of a matrix, which involves the change in primary nucleation, was also reported for the PVDF/PA-6 and PVDF/PBT blends [49].

The studies of iPP/aPS blends [19] have shown that besides the influence of migration phenomenon on the heterogeneous nucleation mode the primary nucleation behavior in those blends is affected also by another phenomenon. This is the ability of the iPP/aPS interface to induce the formation of additional heterogeneous nuclei. This nucleation ability of the iPP-aPS interface is probably connected with the interfacial energy between these polymers, which induces the reasonable decrease of the energy barrier for the formation of heterogeneous nuclei contacting with that interface [19]. Another reason for activity of the interface toward primary nucleation of crystallization of the matrix might be the possibility of the partial orientation of matrix macromolecules at the interfacial layers. According to the theory of Ziabicki [59], such an orientation could induce the lowering of the energy barrier for the formation of critical size primary nucleus. In this way the interfaces present in the blend volume may induce the formation of a new fraction of primary nuclei.

2.3.2 Copolymers

The experimental data concerning crystallization of PP copolymers, although very limited, demonstrate that the primary nucleation behavior in copolymers differs from that in both plain homopolymer and blends. Such difference can be understand easily – the noncrystallizable copolymer units interrupt or terminate crystal growth in the molecular chain direction because they are chemically bonded to crystallizable units and their position along the chain is fixed. Noncrystallizable copolymer units represent defects with various possible disruptive effects. The consequences of their presence on nucleation and growth of crystals depends strongly on their chemical structure, configuration as well as position and concentration along the chain. The numerous theoretical and experimental studies regarding the effect of copolymerization on the crystallization process and properties of resulting crystals were reviewed by Wunderlich [60]. Here, we will only discuss briefly the primary nucleation in copolymers containing isotactic sequences of propylene.

The key feature of chains of a crystallizable copolymer is the order of comonomer units in a copolymer chain, which determines the length of crystallizable sequences and the phase structure of a copolymer melt prior to the crystallization process. In the case of random copolymers there is random distribution of the length of iPP sequences. Moreover, random copolymers rarely exhibit phase separation. On the other hand, in block copolymers the blocks are quite uniform in length and a phase separation in the molten copolymer is frequently observed.

The random length distribution of iPP sequences is random copolymers ensures that only a part of the crystallizable sequences can participate in the formation of primary nuclei; sequences that are too short are excluded from the nucleation process. This decreases the density of primary nucleation as compared to iPP homopolymer. Another important factor limiting the number of primary nuclei in copolymers is the energy barrier due to rejection of noncrystallizable comonomer units out of a forming nucleus, i.e. local phase segregation. This energy barrier is similar in nature to that discussed previously for homogeneous blends, but probably reaches much higher level than in blends because noncrystallizable elements are chemically bonded with segment(s) forming a nucleus. On the other hand, the rejection of noncrystallizable units at the crystal–melt interface chages the interfacial free energies of the forming crystal. This may increase or decrease the energy barrier for critical nucleus formation, depending on the chemical nature of the comonomer. The presence of the comonomer units in the chain also influences the chain mobility, and thus modifies the conditions of the transport of the crystallizing segments across the melt-crystal interface (again, enahancement or reduction, depending on the nature of the comonomer). The experimental data obtained for propylene random copolymers [53, 56] suggest that the net result of the discussed barriers is frequently positive and the primary nucleation in random copolymers is significantly lower than in plain homopolymer or in blends.

Block copolymers have different conditions for nucleation. The crystallizable blocks have similar length, so that practically any of them may participate in the nucleation process. The phase separation in block copolymers usually occurs prior to crystallization from the melt, so there is no additional energy barrier for phase segregation during formation of a nucleus, as in the case of random copolymers. However, similarly to random copolymers one can expect a variation (frequently a significant decrease [60]) of the interfacial free energies of the crystal in a copolymer melt compared to that growing in a parent homopolymer melt, which in turn causes a variation of the energy barrier for critical nucleus formation. Additionally, the interfaces already present in the molten copolymer can play an important role in the nucleation process. Copolymer chains pass across these interfaces. This causes a local orientation of the chains near the interface. Such an orientation effect should strongly enhance the primary nucleation in block copolymers. The above remarks suggest the possibility of both a decrease as well as an increase of the primary nucleation density in block copolymers, as compared to a respective homopolymer, depending on the chemical structure of the noncrystallizable blocks. A probable enhancement of the nucleation in block copolymers is strongly supported by the experimental data obtained by Drzewinski [33], who demonstrated a remarkable increase of the primary nucleation density in propylene–butadiene block copolymer as compared to iPP homopolymer, and an even more significant increase if compared to a blend with a similar propylene content.

2.4 CONCLUDING REMARKS

The results reported in this chapter concerning the primary nucleation behavior in polymer blends show that under normal crystallization conditions the changes in nucleation density can be attributed mainly to the changes in the heterogeneous nucleation mode and in part to the self-seeding nucleation.

In the studies discussed it was pointed out that there is a variety of phenomena inducing the changes of heterogeneous nucleation in the blends as compared to a plain crystallizing component. The behavior of heterogeneous nucleation in several blends is summarized in Table 2.1.

Table 2.1 The phenomena influencing heterogeneous primary nucleation in polypropylene based blends

Blend [a]	*Influence of complete or partial miscibility*	*Migration of impurities*	*Crystallization of second component*	*Influence of the interface*	*Others effects*	*Reference*
iPP/aPP	↘ [b]					[13]
iPP/iPB	↘		↗↗ [c]			[15, 20]
iPP/LDPE		↘↘	↗↗ [c]			[16, 21, 22]
iPP/HDPE		↘↘	↗↗↗			[18, 20–22]
iPP/LLDPE		↘↘				[37]
iPP/EPM		↗↗↗				[17, 28–30]
iPP/EPDM		↗↗↗				[29, 31]
iPP/SBR		↗?			↗?	[28, 31]
iPP/PIB		↗?			↗?	[29, 30]
iPP/IR		↗?			↗?	[28]
iPP/BR		↘↘				[33]
iPP/TOR				↗↗		[36]
iPP/PS		↗↗↗		↗↗		[19]

[a] data concern the crystallization of the polymer mentioned first in the blend code.
[b] ↗, an increase of nucleation density in blend; ↘, a decrease of nucleation density in blend (the number of arrows is related to the intensity of the effect; ? means that the mechanism is suggested in this review, although not evidenced in the cited reference).
[c] found for samples crystallized nonisothermally.

Two groups of factors have a deciding influence on the changes of primary nucleation in polymer blends:

1. The properties of the minor polymer in the blend including: miscibility with the crystallizing component, glass transition temperature, ability to crystallize and the temperature range in which crystallization is possible, surface tension of the melt of this polymer.
2. The method of blend preparation (mixing the components) and the parameters of the resulting phase structure of the blend, including the sizes of inclusions of second component and the total surface area of the interface. In the case of melt-mixing very important parameters of the mixing process appear to be the duration and the intensity of mixing (shear rate). The migration of additives and impurities occurring during the mixing process strongly influences the nucleation behavior, thus the

parameters of the mixing process are especially important. In blends in which dispersed polymer is able to crystallize, the size of its inclusions, controlled by the parameters of the mixing process, is one of the most important factors determining the crystallization kinetics of the blend.

Primary nucleation in crystallizable copolymers was studied less extensively than that in blends and further detailed studies are necessary to describe the nucleation behavior in these systems. It emerges from the existing data that nucleation in copolymers in influenced by many effects, depending strongly on the copolymer composition and the ordering of the units along the chain. It would be possible to control quite precisely the primary nucleation by modifying the chemical structure of the copolymer.

2.5 APPENDIX: SYMBOLS AND ABBREVIATIONS

aPP	Atactic polypropylene
aPS	Atactic polystyrene
BR	Polybutadiene
c	Concentration of crystallizable polyment
DSC	Differential scanning calorimetry
EPDM	Ethylene–propylene diene terpolymer
EPM	Ethylene–propylene random copolymer
HDPE	High density polyethylene
iPb	Isotactic polybutene-1
iPP	Isotactic polypropylene
iPS	Isotactic polystyrene
I_o	Nucleation constant for crystallizable polymer
IR	Polyisoprene
LDPE	Low density polyethylene
LLDPE	Linear low density polyethylene
PBT	Poly(butyleneterephthalate)
PCL	Polycaprolactone
PEO	Poly(ethyleneoxide)
PET	Poly(ethyleneterephthalate)
PIB	Polyisobutylene
PMMA	Poly(methylmethacrylate)
POM	Poly(oxymethylene)
PPO	Poly(phenyleneoxide)
PPS	Poly(phenylenesulfide)
PVC	Poly(vinylchloride)
PVDF	Poly(vinylidenefluoride)
SALS	Small angle light scattering
SBR	Styrene–butadiene rubber

T_a	Annealing temperature
T_g	Glass transition temperature
T_{gb}	Glass transition temperature of the blend
TOR	*trans*-polyoctylene
U^*	Activation energy of viscous flow
$\Delta g'$	Free enthalpy of fusion
Δg_f	Free enthalpy of melting of the crystal phase
Δg_m	Free enthalpy of mixing of blend components

2.6 REFERENCES

1. Paul, D. R. and Newman, S. (eds) (1978) *Polymer Blends*, Academic Press, New York.
2. Martuscelli, E. (1984) *Polymer Engineering and Science*, **24**, 563–86.
3. Friedrich, K. (1978) *Progress in Colloid and Polymer Science*, **64**, 103–12.
4. Wunderlich, B. (1976) *Macromolecular Physics*, Vol. II, Crystal Nucleation, Growth, Annealing, Academic Press, New York, pp. 1–72.
5. Keith, F. J. and Padden, H. D. (1964) *Journal of Applied Physics*, **35**, 1270–85.
6. Keith, F. J. and Padden, H. D. (1964) *Journal of Applied Physics*, **35**, 1286–98.
7. Stein, R. S., Khambatta, F. B., Warren, F. P. *et al.* (1978) *Journal of Polymer Science, Polymer Symposia*, **63**, 313–28.
8. Wenig, W., Karasz, F. E. and MacKnight, W. J. (1975) *Journal of Applied Physics*, **46**, 4194–8.
9. Khambatta, F. B., Warner, F., Russel, T. and Stein, R. S. (1976) *Journal of Polymer Science, Polymer Physics Edition*, **14**, 1391–1424.
10. Martuscelli, E. and Demma, G. (1980) Morphology, crystallization and melting behavior of poly(ethylene oxide)/poly(methyl methacrylate) blends, in *Polymer Blends*, vol. 1, (eds E. Martuscelli, R. Palumbo, M. Kryszewski), Plenum Press, New York, pp. 101–21.
11. Hu, S. R., Kyu, T. and Stein, R. S. (1987) *Journal of Polymer Science, Polymer Physics Edition*, **25**, 71–87.
12. Z. Bartczak, (1988) Primary Nucleation and Growth of Spherulites in the Blends of Polypropylene, PhD Thesis, Polish Academy of Science, Łódź.
13. Bartczak, Z. and Galeski, A. (1990) *Polymer*, **31**, 2027–38.
14. Bartczak, Z. and Martuscelli, E. (1987) *Makromolekulare Chemie*, **188**, 445–53.
15. Bartczak, Z., Galeski, A. and Pracella, M. (1994) *Journal of Applied Polymer Science*, in press.
16. Galeski, A., Bartczak, Z. and Pracella, M. (1984) *Polymer*, **25**, 1323–6.
17. Bartczak, Z., Galeski, A., Martuscelli, E. and Janik, H. (1985) *Polymer*, **26**, 1843–8.
18. Bartczak, Z., Galeski, A. and Pracella, M. (1986) *Polymer*, **27**, 537–43.
19. Bartczak, Z., Galeski, A. and Krasnikova, N. P. (1987) *Polymer*, **28**, 1627–34.
20. Siegmann, A. (1979) *Journal of Applied Polymer Science*, **24**, 2333–45.
21. Martuscelli, E., Pracella, M., Della Volpe, G. and Greco, P. (1984) *Makromekulare Chemie*, **185**, 1041–61.
22. Teh, J. W. (1983) *Journal of Applied Polymer Science*, **28**, 605–18.
23. Lovinger, A. and Williams, M. L. (1980) *Journal of Applied Polymer Science*, **25**, 1703–13.
24. Martuscelli, E., Pracella, M., Avella, M. *et al.* (1980) *Makromolekulare Chemie*, **181**, 957–67.

25. Noel III, O. F. and Carley, J. F. (1984) *Polymer Engineering and Science*, **24**, 488–92.
26. Grebowicz, J. and Pakula, T. (1980) Crystallization and melting of components in blends of polyethylene and polypropylene, in *Polymer Blends*, Vol. 1, (eds E. Martuscelli, R. Palumbo and M. Kryszewski), Plenum Press, New York, pp. 87–100.
27. Rybnikař, F. (1988) *Journal of Macromolecular Science, Physics*, **B27**, 125–44.
28. Karger-Kocsis, J., Kalló, A., Szafner, A. *et al.* (1979) *Polymer*, **20**, 37–43.
29. Martuscelli, E., Silvestre, C. and Abate, G. (1982) *Polymer*, **23**, 229–37.
30. Martuscelli, E., Silvestre, C. and Bianchi, L. (1983) *Polymer*, **24**, 1458–68.
31. Jang, B. Z., Uhlmann, D. R. and Van der Sande, J. B. (1984) *Journal of Applied Polymer Science*, **29**, 4377–96.
32. Kalfoglou, N. K. (1985) *Angewandte Makromolekulare Chemie*, **129**, 103–19.
33. Drzewinski, M. A. (1986) The Structure–Properties of a Model Set of Crystalline Block Copolymer–Homopolymer Blends, DSc Thesis, Massachusetts Institute of Technology.
34. Varga, J. and Garzó, G. (1990) *Angewandte Makromolekulare Chemie*, **180**, 15–33.
35. Chodak, I., Janigova, I. and Romanov, A. (1991) *Makromolekulare Chemie*, **192**, 2791–800.
36. Wenig, W. and Fiedel, H. W. (1991) *Makromolekulare Chemie*, **192**, 191–200.
37. Long, Y., Stachurski, Z. and Shanks, R. A. (1991) *Polymer International*, **26**, 143–6.
38. Pukánszky, B., Tüdős, F., Kalló, A. and Bodor, G. (1989) *Polymer*, **30**, 1399–1406.
39. Martuscelli, E., Riva, F., Selliti, C. and Silvestre, C. (1985) *Polymer*, **26**, 270–82.
40. Kishore, K. and Vasanthakumari, R. (1986) *Polymer*, **27**, 337–43.
41. Nadkarni, V. M. and Jog, J. P. (1986) *Journal of Applied Polymer Science*, **32**, 5817–28.
42. Nadkarni, V. M. and Radhakrishan, S. (1987) *Polymer Engineering Science*, **27**, 547–52.
43. Shingankuli, V. L., Jog, J. P. and Nadkarni, V. M. (1988) *Journal of Applied Polymer Science*, **36**, 335–52.
44. Flexman Jr, E. A. (1985) *Modern Plastics*, **62**, 72–6.
45. Ghijsel, A., Groesbeek, N. and Yip, C. W. (1982) *Polymer*, **23**, 1913–16.
46. Aref-Azar, A., Hay, J. N., Marsden, B. J. and Walker, N. (1980) *Journal of Polymer Science, Polymer Physics Edition*, **18**, 637–43.
47. Baitoul, M., Saint-Guirons, H., Xans, P. and Monge, P. (1981) *European Polymer Journal*, **17**, 1281–88 (in French).
48. Klemmer, N. and Jungnickel, B. J. (1984) *Colloid and Polymer Science*, **262**, 381–6.
49. Frensch, H. and Jungnickel, B. J. (1989) *Colloid Polymer Science*, **267**, 16–27.
50. Kowalewski, T., Ragosta, G., Martuscelli, E. and Galeski, A. (1994) *Polymer*, submitted.
51. Burns, J. R. and Turnbull, D. (1966) *Journal of Applied Physics*, **37**, 4021–6.
52. Koutsky, J. A., Walton, A. G. and Baer, E. (1967) *Journal of Applied Physics*, **38**, 1832–9.
53. Avella, M., Martuscelli, E. and Della Volpe, G. (1986) *Makromekulare Chemie*, **187**, 1927–43.
54. Marigo, A., Morega, C., Zanetti, R. *et al.* (1989) *Makromolekulare Chemie*, **190**, 2805–13.
55. Guidetti, G. P., Busi, P. and Giulianelli, I. (1983) *European Polymer Journal*, **19**, 757–9.

56. Crispino, L., Martuscelli, E. and Pracella, M. (1980) *Makromolekulare Chemie*, **181**, 1747–55.
57. Blundell, D. J. and Keller, A. (1968) *Journal of Polymer Science, Part B*, **6**, 433–40.
58. Rostami, S. (1990) *Polymer*, **31**, 899–904.
59. Ziabicki, A. (1977) *Journal of Chemical Physics*, **66**, 1638–43.
60. Wunderlich, B. (1976) *Macromolecular Physics*, Vol. II, Crystal Nucleation, Growth, Annealing, Academic Press, New York, pp. 259–71.

3

Polypropylene alloys and blends with thermoplastics

L. A. Utracki and M. M. Dumoulin

3.1 INTRODUCTION TO BLENDS

The following definitions are adopted:

1. *Polymer blends* are mixtures of at least two macromolecular species; they are either *miscible* or *immiscible*, where these terms are understood in a purely thermodynamic sense (that is, in miscible blends the level of miscibility extends to the molecular level).
2. *Polymer alloys* are immiscible, compatibilized blends with modified interphase and morphology.
3. *Compatibilization* is a process of modification of the interphase and/or morphology by either physical or chemical means. A *compatibilized blend* has been modified in such a way. The term *compatible blend* has been badly misused to indicate either a miscible or immiscible system, and it should be avoided.

Polymer alloys and blends (PABs) are increasingly important industrial materials. In 1990, they claimed over 30% of the polymer resin market and with an annual growth of some 9.6% (four times the growth rate of the plastics industry), their role can only increase. Currently, 57% of engineering resins are blended. By 1994, the North American production of PABs will reach one million tons, worth about US$5 billion. By the end of this millennium, 90% of all polymers will be used either in blends or in composites. The principal consumer of PABs is the transportation industry. In 1990, 46% of PABs were manufactured in North America, 34% in Europe, and 20% in East Asia, the fastest growing region.

Polypropylene: Structure, blends and composites. Edited by J. Karger-Kocsis.
Published in 1995 by Chapman & Hall, London. ISBN 0 412 58430 1

The physical properties of PABs can be altered to satisfy a wide range of applications. A desirable performance can be achieved by proper selection of blend ingredients, followed by control of morphology via appropriate methods of compatibilization, compounding and processing [1–5]. Blending is known to improve the impact strength, mechanical properties, and chemical and solvent resistance, to enhance processability, abrasion resistance, flame retardancy, etc. Analysis of recent patents indicates that processability is becoming one of the most important criteria. This is understandable as the emphasis shifts to high performance, difficult to process specialty resins. For these the processing temperature, T_p, is near or even above the thermal degradation temperature, $T_d \leqslant T_p$. Since blending can lower T_p by as much as 60 °C, it frequently enables safe processing.

From the commercial point of view, isotactic polypropylene (PP) blends with polyethylene (PE), with ethylene–propylene copolymer (EPR) or terpolymer (EPDM) (containing a non-conjugated diene), and with polyamide (PA) are of particular importance. They are manufactured in large quantities and have been growing at superior rates. In the following text, these three types of blends will be discussed in much greater detail than the mixtures with polycarbonate (PC), styrenics, thermoplastic polyesters (PEST), and other polymers.

3.1.1 Miscibility

Owing to the negligibly small entropy of mixing, the great majority of polymer blends are immiscible [1–6]. The only PP blends reported miscible are those with atactic polypropylene, (aPP), with EPR containing $w \leqslant 10\%$ of ethylene, and with polybutene-1 (PB-1) [7, 8]. Even in these cases, the miscibility is limited to the amorphous phase as PP crystallizes out from the mixtures.

For most polymer blends, the Huggins–Flory binary thermodynamic interaction parameter is large, $\chi > 0$, engendering low degree of molecular entanglement in the interphasial region. As a result, the interfacial adhesion in both molten and solid states is usually poor. For example, the value of the interphasial viscosity was estimated to be about 100 times smaller than that of the neat resins [9]. For uncompatibilized blends, the solid state properties (ultimate elongation) versus composition usually show a strong negative deviation from additivity. Thus, preparation of commercially viable blends must involve compatibilization, by means of compatibilizer addition or reactive blending.

Compatibilization of PP blends is based on either copolymerization or grafting of PP. In both cases, one part of the resulting compatibilizer will be miscible with the PP phase, the other with the second polymeric component. For example, EPR is used for compatibilization of PP with PE. The propylene part of EPR provides miscibility with PP and the ethylene part does so with PE. Acid or anhydride modified PP, (m-PP) is used for blends with PA, PC or PEST. There are a few acidic or anhydride groups per m-PP macromolecule. These react chemically with amide or ester groups of PA, PC or PEST, forming the compatibilizing graft copolymer. In such a copolymer, the PP part provides miscibility with the PP phase and the grafted part does so with the other polymeric component. It is interesting that this kind of compatibilization has very little effect on the glass transition temperature T_g of both ingredients.

Once the blend components are selected, the properties can be controlled by morphology (which in turn depends on the molecular parameters of the blend components) and composition, as well as by the compounding and processing methods. In the case of amorphous blends, the morphology is defined by the size and shape of the two phases, their distribution and orientation. In the following text, this type of morphology will be referred to as *macro-morphology*. In blends of semicrystalline polymers, such as PP, blending affects the crystallinity, the study of which will be referred to as *micro-morphology*. Both macro- and micro-morphology depend on the thermodynamic and rheological properties of the ingredients and the methods of compatibilization, as well as on the deformation and thermal histories.

3.1.2 Macro-morphology

The macro-morphology of polymer blends describes the form and size of the macromolecular phases formed during compounding or blending. At low volume fraction of the dispersed phase, $\phi < \phi_p$ (in three dimensions the percolation threshold, $\phi_p \approx 0.156$), the dispersed morphology is best described by the micro-rheology [1–5, 10].

The micro-rheology identifies two parameters controlling the ability of one liquid to disperse in another: the viscosity ratio, λ, and the capillarity number, κ:

$$\lambda = \eta_1/\eta_2 \qquad \text{and} \qquad \kappa = \sigma_{12} d/\nu_{12} \tag{1}$$

where η_1 and η_2 represent viscosity (determined at the shear stress, σ_{12}) of the dispersed and continuous phase, respectively, d is the initial drop diameter, and ν_{12} is the interfacial tension coefficient. In the shear field, the Newtonian drop in a Newtonian liquid remains unbroken for $\lambda > 3.8$, or $\kappa < 0.1$. For $1 < \kappa < 4$ the drop elongates and breaks, while for $\kappa > 4$ the drop deforms affinely with the matrix and breaks only when κ (that is, σ_{12}) is reduced to $\kappa < 4$.

Another important but less frequently discussed parameter is the time for deformation, t_d, and breakup, t_b, expressed in a dimensionless form as:

$$t_i^* = 2t_i \dot{\gamma} \nu_{12}/\sigma d = 2\gamma_i/\kappa, \qquad \text{where} \qquad i = d, b. \tag{2}$$

Thus in shear flow, for these processes to occur the following conditions must be satisfied:

Deformation	$0.1 < \kappa_d < 1.0$ and	$t_d^* > 25$
Breakup	$\lambda < 3.8$, $\kappa_b > 1$ and	$t_b^* > 160$

In viscoelastic liquids, a parallel behavior has been observed. However, here the values of critical parameters depend on the normal stresses, N_1, i.e. on the system elasticity. The elasticity facilitated the initial drop deformation but made breakup more difficult.

As the concentration increases toward the phase inversion level, $\phi_{Ii} > \phi_p$, the co-continuous structures are developing. For emulsions and blends, the conditions for phase inversion are given by [11]:

$$\lambda = [(\phi_m - \phi_{I2})/(\phi_m - \phi_{I1})]^{[\eta]\phi_m}; \qquad \phi_m = \phi_p \tag{3}$$

where $[\eta]$ is the intrinsic viscosity and $\phi_{I1} + \phi_{I2} = 1$. For the viscosity ratio $\lambda = 1$, Equation 3 predicts $\phi_{I1} = \phi_{I2} = 0.5$, whereas for $\lambda \to \infty$ the phase inversion concentration coincides with the percolation threshold, $\phi_{I1} = \phi_p$. Since ϕ_p depends on system dimensionality, and for two dimensions $\phi_p = 0.02$, the relation correctly predicts the preferential formation of the co-continuous structure in extensional flow.

There are several methods of determining blend morphology. These include optical microscopy (OM), scanning electron microscopy (SEM) and transmission electron microscopy, as well as light, X-ray and neutron scattering techniques. In all cases, specimen preparation is critical. Cryo-fracture, microtoming, and Soxhlet extraction are the methods most frequently used. The techniques vary depending on the polymer type, degree of dispersion and the kind of structure generated during blending (lamellae, droplets, etc.). For electron microscopy, the specimens are usually ultramicrotomed then stained with Br_2, RuO_4 and/or OsO_4, or extracted with a mild solvent, treated with solvent vapors or plasma-eroded with atomic oxygen [12]. For SEM, solvent etching (e.g., with n-heptane) or OsO_4 staining gave good results [13]. Metallization with Au or Pd/Au is required, unless the instrument is operated at low accelerating voltage (0.1 to 2 kV). For TEM, solvent or plasma etching and carbon replication have been used. Analysis of the micrographs requires evaluation of at least 250 particles, thus an image analyzer is recommended. At low concentration, $\phi < \phi_p$, the droplet size distribution usually follows the log-normal distribution [14].

3.1.3 Flow behavior

The multiphase nature of PABs implies that flow responses may be complex. Furthermore, since the rheology depends on the macro-morphology, which in turn is determined by the intensity and type of flow field, the data from different instruments usually do not agree. Continuum mechanics predicts that at low deformation rates, steady state shear, dynamic, and extensional viscosities are all related. However, this is rarely true for PABs. In multiphase blends, the basic assumptions of continuum mechanics are not valid. Indeed, owing to flow modification of blend morphology, in most cases the three sets of data are quite different.

For concentrated PABs and/or systems with high viscosity ratios, $\lambda > 1$, an apparent yield stress is usually observed. However, since the gel-like structure originates not from the covalent bonding but from the inter-drop physical interactions, the yield stress has a dynamic character [2]:

$$\sigma_y(\dot{\gamma}) = \sigma_y^{\infty}[1 - \exp\{-\tau_y \dot{\gamma}\}]^u = \sigma_y^{\infty}[1 - \exp\{-1/N_D\}]^u \tag{4}$$

where σ_y^{∞} is the yield stress at high deformation rate, τ_y, is the characteristic time for particles aggregation and $u = 0.2$ to 1 is an exponent characterizing polydispersity of the aggregate size. Thus, the magnitude of the apparent yield stress depends on the Deborah number: $N_D \equiv$(test time scale)/(material time scale)$=(1/\dot{\gamma})/\tau_y$. For large values of N_D, σ_y vanishes, whereas for $N_D \ll 1$ the material behaves like a solid, with the yield stress $\sigma_y = \sigma_y^{\infty}$. For strongly interacting, nonlinear viscoelastic systems with yield (for example, such as highly concentrated filled systems), the parameters of Equation 4 depend on strain.

There is a large body of literature dedicated to discussions of the concentration dependence of viscosity, η versus ϕ. The log-additivity rule is a convenient baseline:

$$\log \eta = \sum_i w_i \log \eta_i \tag{5}$$

Here η_i represents the constant stress viscosity of ingredient i present at weight fraction w_i. Very few systems follow Equation 5; about 60% show a positive deviation (PDB), about 30% a negative deviation (NDB), and the rest a mixed one (PNDB). It was recently postulated that the flow of PABs is controlled by two mechanisms: (i) the emulsion-like behavior causing viscosity to increase above the log-additivity rule, and (ii) the interlayer slip, responsible for the negative deviation [11]. Superposition of these two mechanisms is responsible for the diversity of rheological responses. The postulate can be expressed in the form:

$$\log \eta = \log \eta_L + \Delta \log \eta^E \tag{6}$$

where the interlayer slip effect is introduced using Lin's equation [15]:

$$\log \eta_L = -\log[1 + \beta(\phi_1 \phi_2)^{1/2}] - \log(\phi_1/\eta_1 + \phi_2/\eta_2) \tag{7}$$

and the emulsion-like contribution by:

$$\Delta \log \eta^E = \eta_{max}\{1 - [(\phi_1 - \phi_{I1})^2/(\phi_1 \phi_{I2}^2 + \phi_2 \phi_{I1}^2)]\} \tag{8}$$

In Equations 7 and 8, β is the interlayer slip coefficient and η_{max} the parameter quantifying the emulsion-like behavior. Equation 6 was found to be useful in determining the flow mechanisms in several PABs.

3.1.4 Micro-morphology

The fine structure in PP blends is frequently studied under SEM, using specimens etched with chromic acid, saturated solution of CrO_3 in oleum, aromatic and chlorinated solvents, or an electronic beam. Recenty, the use of toluene/xylene vapors at the boiling point, followed by short immersion, was found to be particularly successful in removing the amorphous phase [16]. X-ray, neutron or light diffraction methods also provide valuable information.

In PABs, the presence of another polymeric ingredient affects crystallinity. It is useful to consider three situations: simultaneous crystallization of both components, crystallization of one component in the presence of a melt, and crystallization of one component in the presence of the second component in the solid state [17].

1. Simultaneous crystallization takes place when ranges of the crystallization temperature, T_c, of two polymers overlap and their crystallization rates are similar. Here, crystallization frequently leads to mixed spherulites or axialites without change of the unit cell parameters. Depending on the compactness of lamellae the physical properties of PAB may be either additive (most common) or show positive or negative deviation.
2. Most often, crystallization occurs in the presence of the liquid phase of the second component – amorphous or semicrystalline. For such systems, Paul and Barlow [18] predicted five crystallization patterns: (i) no effect, (ii) retardation, (iii) delay, (iv) acceleration, and (v) causing crystallization of a polymer unable to crystallize by itself. Crystallization of PP with PE, PB-1 or polycaprolactone (PCL) follows these patterns.
3. Crystallization in the presence of crystals of the other component takes place through a heterogeneous nucleation. Since nucleation is usually the rate controlling step and the rate of crystallization for heterogeneous nucleation is usually higher than for the homogeneous one, here the crystallization rate is expected to be significantly higher than that of neat resins. The enhanced nucleation also tends to increase the crystallization temperature, T_c. Crystallization of PP in the presence of PA or polyvinylidenefluoride (PVDF) exemplifies this behavior.

3.1.5 Mechanical properties

It is convenient to discuss the mechanical properties of any material in terms of two deformation ranges: the low strain behavior (e.g. moduli, strain recovery, etc.) and the ultimate properties generally described by fracture mechanics.

In the simplest case, the modulus of a two component system depends on moduli of both ingredients, as well as on concentration and anisometry of the dispersed phase. Takayanagi, Uemura and Minami [19] considered such a system. Their relation, written for the relaxation and complex moduli, can be expressed as:

$$E(t, T) = (1-\alpha) E_2 + \alpha/[(1-\delta)/E_2 + \delta/E_1]$$
$$\left.\begin{aligned} E' &= (1-\alpha) E'_2 + \alpha X/(X^2+Y^2) \\ E'' &= (1-\alpha) E''_2 + \alpha Y/(X^2+Y^2) \end{aligned}\right\} \qquad (9)$$

where $X \equiv (1-\delta) E'_2/E_2^{**2} + \delta E'_1/E_1^{**2}$; $Y \equiv (1-\delta) E''_2/E_2^{**2} + \delta E''_1/E_1^{**2}$; $E^2 = E'^2 + E''^2$ and $E_i^{**} \equiv E_i'^2 + E_i''^2$. α and δ are the model parameters whose product $\alpha\delta = \phi$. The Takayanagi relations provide adequate description for oriented, isothermal systems with α and δ values giving information on the morphology. Recent analysis of PP/PA data indicates that, contrary to the theoretical assumptions, α and δ values (calculated from data below the glass transition temperatures, $T < T_g$) depend on T [20]. In general, the time–temperature superposition principle, t–T, is not applicable to PABs [21, 22].

The ultimate properties are determined using either impact (Charpy or Izod) or low speed tensile methods. In the latter case, for semicrystalline polymers and their blends, a purely brittle fracture has usually been observed. Thus, the critical strain energy release rate, G_c, may be determined following Williams' postulate [23, 24] that the energy observed at fracture, U, is a linear function of the product. BDF, where B is the specimen thickness, D its depth and F a tabulated parameter. For example, the computed values of G_c for PA-6,6, a supertough PA-6,6, and for a compatibilized PP/PA-6,6 blend (measured in parallel and perpendicular to the extrusion direction) are, respectively, $G_c = 8.2$, 75.7, 10 and 5.8 kJ/m^2 [20].

The compounders and processors of polymer blends must recognize that PABs performance depends not only on parameters controlling the homopolymer systems (such as molecular weights (MW), their distribution (MWD), thermal and stress histories) but also on the morphology. Several books and reviews have been published on PABs science and technology [2, 6, 25–40]. Readers interested in blend technology are strongly encouraged to consult these references. There are also several reviews on rheology, processing and properties of PP blends [1, 17, 22, 41–45].

3.2 BLENDS WITH POLYETHYLENES

3.2.1 Introduction

Owing to interesting properties and low cost, PE constitutes the largest group of thermoplastics. Consequently, their blends with PP have attracted much commercial interest [46–54]. One of the reasons for adding PE to PP is to improve the low temperature impact behavior of the latter polymer. The PE/PP blends are immiscible. Marked difference in properties between PP blends containing different PEs have been reported.

Compatibilization of PP/PE blends can be accomplished by addition of a copolymer, EPR or EPDM, by reactive blending, or by post-blending treatment, using chemical crosslinking, electron beam or γ-radiation. The reactive compatibilization is usually accomplished by addition of a peroxide. The reaction leads to degradation of PP and branching or crosslinking of PE. Optimization, then tight control of the process is required [55]. The effects of blending and γ-irradiation on PP/low density PE (LDPE) performance has been reported [32, 56, 57]. Both reactive compatibilization and irradiation led to lower blend crystallinity. Thus, the micro-morphological character of PP/PE blends is controlled by composition, presence of nucleating agents, type of compatibilizer and thermal or radiative treatment [17].

Most of the PP/PE blends contain the traditional LDPE, linear LDPE (LLDPE) or high density PE (HDPE). However, blending with ultralow density PE, (ULDPE) ($\rho = 890\,kg/m^3$), and ultrahigh MW HDPE (MW $\geqslant$ 1000 kg/mol) (UHMWPE) is also known [16, 58, 59].

3.2.2 Flow properties

Plochocki coined the term '*rheologically particular composition*' (RPC) for concentrations at which an extremum is observed in a viscoelastic function [60]. He discussed several polyolefin blends for which the property–composition curves departed from linearity, and noted that this type of behavior is caused by density variations at the interface. Near an RPC, the components diffuse into each other, causing density fluctuations and changes in the size of the domains. Thus, an RPC provides evidence for the morphological changes that affect the flow as well as the viscoelastic and mechanical responses. The presence of RPCs was reported for most PP/PE blends in concentration dependence of viscosity, η, or the first normal stresses' difference, $N_1 = \sigma_{11} - \sigma_{22}$ [61].

The capillary viscosity, η, versus concentration, ϕ, dependence for three PP resins blended with HDPE at $\lambda = 5$, 1 and 0.5 was reported by Valenza, La Mantia and Acierno [62]. The relationship was found to be additive (see Equation 5) with the exception of the isoviscous system, $\lambda = 1$, where slight NDB was observed. In another publication [63], the η versus ϕ dependence was found to show either NDB (it followed Equation 7 with $\beta = 1$) or PNDB, while the entrance pressure drop (Bagley correction) showed a positive deviation. Different blending procedures and treatments did not affect the results. The photomicrographs showed an annular structure of the extrudates. The capillary flow of HDPE/PP blends was also studied by

Alle *et al.* [64, 65]. For $T = 180–210\,°C$, the flow curves, η versus $\dot{\gamma}$, superimposed on the time–temperature master curve with η_0 following an Arrhenius dependence. The η versus ϕ plots showed NDB. For homopolymers and their blend containing 75% PP, the extrudate swell, B, appeared to be temperature-independent, but highly temperature-sensitive for the other compositions. The latter observation led to the conclusion that phase inversion occurred at ϕ_I (PP)$\approx$0.75.

Recently Fujiyama and Kawasaki [66, 67] reported on the rheological behavior of three PP resins blended with three HDPE ones. Capillary and dynamic test methods were used. At constant shear rate, the capillary η versus ϕ plots showed NDB. The extrudate swell was found to be larger than that for homopolymers and unrelated to elasticity of the system. The critical shear stress for melt fracture changed with PE content from $\sigma_c = 120–620\,kPa$. The drop dispersion was observed up to about $w = 30\,wt\%$; the finest for $\lambda \approx 1$. At low frequencies, $\omega \leqslant 10^{-2}$ (rad/s), the dynamic viscosity, η', and storage modulus, G', versus w(PE) plots showed either PDB or PNDB, while at constant loss modulus level, $G'' = 100$ Pa, the G' versus w(PE) plot showed only PDB. A time-dependent yield stress (see Equation 4) was also observed. Melt rheology and strain induced crystallization in PP/HDPE blends was discussed recently by Chuu and Tsai [68].

The capillary flow of PP/LDPE blend with $\lambda \approx 1$ was examined by Santamaria *et al.* [69]. They reported that extrudate shrinkage was composition-dependent with a maximum occurring near ϕ_I(PP)$\approx$0.75. The co-continuous morphology was responsible for the largest extrudate swell. The recent data on flow of PP/ULDPE showed similar dependencies as those observed for blends with LDPE. The concentration dependence of η^* indicated predominance of the emulsion-like effects near ϕ_I with the interlayer slip evident in all other regions [58].

Dumoulin, Farha and Utracki [70] reported that for PP/LLDPE = 50/50 blend the capillary viscosity followed Equation 5. Addition of EPR did not cause any significant change in either the viscosity or mechanical properties (at room temperature or at $-40\,°C$), but it eliminated the interlayer slip in capillary flow, and affected the entrance–exit pressure drop or the Bagley correction. The extensional and dynamic shear flow were also examined. For neat resins, the Newtonian plateau at low test frequencies was nearly reached, but for the blends, with or without EPR, there was an indication of apparent yield stress. The stress growth functions in shear and in uniaxial extension were measured for pure polymers and the blends; a good agreement between the two types of deformation as obtained and no strain hardening was observed.

A more detailed study on rheology of two PP resins blended with one LLDPE material ($\lambda = 1.7$ and 0.25) was later reported by Dumoulin *et al.* [21, 71, 72]. It was found that during compounding with LLDPE, the PP degraded significantly. The shear induced degradation was apparently catalyzed by either LLDPE, or its impurities, primarily affecting the highest MW fractions of PP. The dynamic moduli, G' and G'', indicated a yield stress at the mid-range concentrations where a co-continuous morphology was observed. The dynamic test data were fitted to a four-parameter modified Carreau equation from which the gross frequency relaxation spectra were computed. The concentration dependence of the zero shear viscosity, η_0, was PNDB type. These data were later analyzed using Equations 6–8. The analysis indicates that both mechanisms were equally important; the presence of a large interlayer slip provided evidence of poor thermodynamic interactions between PP and PE. It may be of interest to note that similar analysis of LDPE/LLDPE gave a smaller value of β, indicating better miscibility [11].

3.2.3 Morphology

PP/PE morphology has been discussed recently [58, 68, 73]. The macromorphology of PP blends shows the characteristic features of PABs: dispersed droplets at $\phi \leqslant \phi_p$, mixed dispersed forms at $\phi_p < \phi < \phi_I$, and co-continuous structures near the phase inversion $\phi \approx \phi_I$.

Blends of PP containing up to 40 wt% PE were examined under OM by Wenig and Meyer [74]. Within this ϕ range, the PE phase was made of almost globular drops, a few micrometers in size. Their size and size distribution increased with increasing temperature and annealing time. The authors reported that PE drops were occluded within the growing PP spherulites and the growth rate of the PP spherulites was ϕ-independent.

A number of studies on the micro-morphology of PE/PP blend have been reported. For example, the effects of high strain on the micro structure of PP/HDPE blends were discussed by Kuptsov *et al.* [75]. A number of studies examined the crystallization kinetics of polyolefin blends [76–83]. The effect of PE on crystallization of PP has been examined in isothermal experiments at temperatures either below or above T_c(PE). Bartczak, Galeski and Martuscelli [76] studied the kinetics of crystallization of PP blended with either LDPE or with an elastomer. The authors discussed the spherulitic growth at $T > T_c$(PE) in the presence of the dispersed melt and the work required for its rejection or deformation. The experiments indicated that the interfacial energy, the viscosity of the melt, the spherulite growth rate, and the size of the dispersed particles all play a role in the process. In blends, the PP spherulite growth rate was found to be lower than in a homopolymer and its value decreased with an increase of the dispersion degree.

Spherulitic growth at $T > T_c$(PE) in PP/HDPE [77] and PP/LDPE blends [78, 79] has been studied. In contrast with the results in [76], it was observed that the PP spherulitic growth rate is only temperature-dependent; it remains unaffected by the presence of another polymer.

During crystallization of PP, the domains of PE were occluded by the growing crystallization front, and thus incorporated practically undeformed, into the spherulites. Therefore, the crystallization front had to grow around the PE occlusions, creating a weld line where the two fronts met. During the fracture testing these weld lines initiated crazing, allowing for better energy dissipation [78].

Another mechanism responsible for improved mechanical performance of PP/PE blends was postulated by Bartczak and Galeski [80]. While keeping PE molten, the authors used optical and electron microscopy to examine the interface between the growing spherulite and the melt. It was observed that the solidification process involved volume changes, creating voids within the spherulites. The molten PE phase tended to move into these voids, thus increasing the interfacial surface, improving the cohesion and, consequently, enhancing the blend performance. The crystallization of PP at $T > T_c(\text{PE})$ reduced the number of spherulites and increased their size. On the other hand, when the crystallization was carried out at $T \leqslant T_c(\text{PE}) = 127\,°\text{C}$, the PE crystals were found to nucleate PP, increasing the number of spherulites and reducing their size. The finer micro-morphology again resulted in improvement of the overall mechanical performance.

Noel and Carley [81, 82] observed crystallization of an HDPE/PP blend under OM. The onset of crystallization occurred at 134 and 126 °C for PP and PE, respectively, resulting in two distinct crystalline phases. Quenching reduced the mean spherulite diameter, whereas slow cooling caused distortions of the PP spherulites. The authors also studied PP spherulite growth at $T > T_c(\text{PE})$. In agreement with observations reported by Bartczak and Galeski [80], incorporation of HDPE within the growing PP spherulites was reported. The occlusions of HDPE domains caused shape distortion of the PP spherulites. In a similar study [83], a large difference in spherulite size of HDPE and PP was reported: 10 and 100 μm, respectively. An addition of as little as 10% of PE induced a sharp reduction of the PP spherulite size. The SEM showed that at 50/50 composition, the HDPE particles (with diameter $d \approx 2$–$10\,\mu\text{m}$) were dispersed in the PP matrix. Further increase of PP concentration caused a reduction in the HDPE drop size, while its decrease resulted in formation of a co-continuous structure. The reduction of PP spherulite size along with the profusion of intercrystalline links was proposed as the principal mechanism responsible for improvement of the mechanical behavior.

Nakafuku studied the crystallization of PE/PP blends under high pressures [84]. Microscopic examination of these blends suggested absence of interdiffusion of the two polymers. Some occlusion of PE into the PP crystalline phase was again reported.

Owing to the commercial interest in ultradrawn PE fibers and films, there is significant activity in study of the morphology and mechanical properties of the PP/UHMWPE blend. The blends are prepared from a dilute solution through the gel crystallization–drawing process. The stretched film birefringence at a draw ratio DR = 40 was found to be temperature-independent up to $T = 160\,°C$, then showed only a small decrease at $T \leqslant 190\,°C$ [59]. Thus, because of the dense arrangement of the adjacent oriented lamellae of PP and PE crystals, the molecular motion was severely restricted even at these high temperatures (see also [85]). Upon blending, the microhardness of PE and PP lamellae decreased, while the total crystallinity fell below the value estimated from the additivity rule [86].

The fine structure of the ternary PP/EPDM/HDPE blends was studied using TEM after ruthenium staining [12]. The effectiveness of the method was assessed by comparing structures of the blends prepared by way of two processing routes: extrusion and injection molding or two roll milling and compression molding. The first method yielded relatively small, irregularly shaped droplets with EPDM and HDPE randomly distributed throughout the PP matrix. The second method resulted in large shell (EPDM)–core (PE) spherical drops. Here, in systems with HDPE as the matrix, the PP lamellae were seen entering the EPDM dispersed phase, while when HDPE was the dispersed phase, PE lamellae entered the PP matrix. On the other hand, Mao, Zhang and Ying [16] studied the morphology of molded samples of PP and PP/HDPE blends by etching the surfaces with toluene/xylene vapors. The presence of PE did not affect the crystal form of PP.

Morphological observations on LLDPE/PP blends are more scarce. Yeh and Birley [87] examined the thermal and mechanical behavior of PP blends with 5 and 10 wt% LLDPE. The calorimetry showed well-defined melting and crystallization peaks for each polymer. The melting temperature, T_m, values corresponded to those of the neat resins, suggesting that the crystal structure remained unaffected, while T_c(PP) increased by 4 °C, indicating enhanced nucleation by LLDPE. The blend impact properties were significantly better than those of PP. By contrast with the above findings, Long, Stachurski and Shanks [88] reported that T_c(PP) decreased upon addition of LLDPE. The crystallization rate of PP was found to be independent of composition, and the overall crystallinity decreased due to reduction in the density of the primary nuclei. Recently, Gunderson and Chilcote reported on PP blends with LDPE and LLDPE [89]. The blend morphology was studied using OM and differential scanning calorimetry (DSC), while the film performance was evaluated in tensile tests. A phase co-continuity morphology was reported for the 50:50 composition. Near this concentration, PP spherulites were quite irregular in shape. The effect was most likely caused by competition of the two polymers to form a continuous matrix. The solid state was reported to be a mixture of occlusions of the minor phase spherulites within the major phase ones. Tensile strength versus composition plots for the compression molded samples showed a negative deviation from linearity, whereas those of the cast film samples showed a positive deviation. The difference in performance of the PP/LDPE and PP/LLDPE blends was rather small.

The influence of processing on morphology of PP/LLDPE blends was assessed by comparing SEM observations of two blend systems (two PP resins blended with a single LLDPE material; $\lambda = 1.7$ and 0.25) over the whole concentration range after compounding and after injection molding [21, 44]. A discrepancy between the two morphologies was explained by partial solubilization of the dispersed phase induced by high strains and pressures encountered by the material during injection molding. Annealing the injection molded samples at 170 °C caused a large number of small particles to precipitate from the matrix. Co-continuous structures were observed for $w(\text{LLDPE}) = 50$ and 75 wt%, bracketing the phase inversion region. These morphological observations were related to changes in the relaxation time spectrum and occurrence of a yield stress in dynamic flow. It should be noted that the phase inversion region covered the same concentration range for the two systems even though the viscosity ratios were quite different.

Recently the influence of compatibilizer addition at $w = 10$ wt% on the micro-morphology of PP/LLDPE = 4/1 blend was investigated by Flaris, Wasiak and Wenig [90]. EPR and two styrene–ethylene–butylene–styrene triblock copolymers (SEBS) were used as compatibilizers. Upon the addition of the copolymers, the nucleation density as well as the maximum rate of crystal growth increased. The Avrami exponent decreased from $n(\text{PP}) = 2.3$ to $n(\text{PP/LLDPE/SEBS}) = 0.7$, indicating athermal crystallization. On the other hand, the interfacial free energy of the lamellar interfaces was not affected.

3.2.4 Mechanical properties

Blending of PP with PE engenders two competing effects. First, the presence of two immiscible polymers creates a two-phase structure that may be detrimental to blend performance, especially at large strains. Second, the crystallinity of both polymers is affected, which (as discussed above) generally leads to positive effects. Since moduli are measured at small deformations, the effect of immiscibility is small. On the other hand, the properties measured at large strains usually show the impact of the two-phase structure. In addition, compounding plays a significant role in blend performance. For example, when blending PP with PE, an increase in mixing time or its intensity not only improves the degree of dispersion but also increases the thermal and/or mechanical degradation [20, 43, 44, 91, 92].

Owing to immiscibility in PP/PE blends, at large strains one may expect poor ultimate properties. Ethylene–propylene copolymers are the obvious compatibilizers. Their addition has often been reported to be advantageous

[93, 94]. However, in polyolefins, MW and MWD have profound effects on blend performance. This was demonstrated by Dumoulin, Farha and Utracki [70], who showed that the tensile and flexural moduli as well as the tensile yield stress of the noncompatibilized LLDPE/PP = 50/50 blend followed the linear additivity rule. The strain at break of the blend was found to be even higher than that of the more ductile blend component, LLDPE. Addition of EPR only slightly improved these properties, but at the same time it reduced ductility. Overall, the two-component blend was more advantageous. Its properties were easily tuned between these of the more ductile PE and the stiffer PP. These observations are in qualitative agreement with an earlier publication by Ramsteiner *et al.* [95], where the improvement of low temperature impact strength of PP accomplished by addition of PE is discussed. Microscopic investigations showed that dispersed PE particles enhanced the craze initiation and dissipation of impact energy.

Synergistic effects were reported for two-component, non-compatibilized PP/PE blends, e.g. in modulus or tensile strength [81, 82, 96]. For PP/HDPE blends, the relationship between modulus and composition was found to be either linear [96, 97] or PNB type [81, 82, 98]. Since the presence of one component affects the morphology of the other, the properties strongly depend on the polymer characteristics, method of compounding and processing as well as on the total thermal history. For example, it was noted that samples prepared by compression molding were less ductile than those manufactured by injection molding [99]. Furthermore, presence of weld lines markedly reduced the ultimate behavior of the latter specimens. Addition of a small amount of HDPE was found to have a beneficial effect on the drawability of PP fibers [100].

Robertson and Paul [97] observed an almost linear relation between the modulus (or tensile stress) and composition for PP blends with either HDPE or LDPE. However, in both cases, the strain at break showed a strong negative deviation. The mechanical properties of highly oriented PP/PE 50/50 tapes were found to follow a simple parallel model, in which the PP and PE lamellae acted independently of each other [86]. Lovinger and Williams [83] have noted a large positive deviation for the modulus of HDPE/PP blends but also NDB for the ultimate strain. The mechanical properties of unoriented and oriented by cold drawing PP/HDPE blends were investigated by Greco *et al.* [101], who stressed the importance of the molecular parameters for interpretation of blend behavior. Two other works showed a maximum in the E versus w(PP) curve: at w(PP) = 90 wt% [81, 82] and w(PP) = 75 wt% [98]. In the latter work, a mild PDB in the E versus w(PP) dependence was observed, while the strain at yield and break displayed a spectacular drop. Deanin and D'Isidoro [96] related these observations to the components' MW; the poor ultimate properties were associated with weak interfacial interactions. Addition of 10–30 wt% of properly selected EPR improved the performance.

The necking behavior of LDPE/PP blends was studied by Rizzo and Spadaro [102]. It was observed that the ultimate tensile strain was related to the mixing time. A thoroughly mixed blend showed neck formation and drawing, which resulted in large values of the strain at break, while the same composition mixed for 2 min fractured at low strains. Examination of these two blends under SEM showed very different morphologies; the large LDPE islands that could be seen in the specimens mixed for 2 min were invisible in those mixed for 15 min. The latter morphology allowed for better stress dissipation and transfer across the phase boundaries, thus improving the ultimate properties. Barlow and Paul [103] attempted to relate the macroscopic degree of adhesion between the polymer pairs to the performance of their blends. The authors highlighted the lack of adhesion between LDPE and PP and the resulting poor performance.

3.3 BLENDS WITH OTHER POLYOLEFINS

3.3.1 Introduction

Among mixtures with other polyolefins, blends of PP with EPR or EPDM elastomers constitute the largest and most important group [104, 105]. At low concentrations these materials have been used for toughening PP, while at higher concentrations, $w(\text{elastomer}) \geqslant 25$ wt%, they are part of the increasingly popular olefinic thermoplastic elastomers (TPO). For additional enhancement of properties (e.g. notched Izod impact strength), these materials can be compatibilized by grafting PP onto a backbone of EPR [106–109] or by partially crosslinking them [110]. TPOs are manufactured not only by the large resin producers under well-recognized names (e.g. Alcryn, Hi-Fax, Hostalen, Kelburon, Novolen, Pro-Fax, Royalene, Santoprene, SPX, Thermolan, Vestolen) but also by the compounding houses (e.g. ETA-Polymer, FerroFlex, Polytrope, RPI, Telcar, TPE, TPO900). There are over 30 TPO manufacturers in North America alone. In the US consumption of TPOs in 1991 was 63 kton. The current world production of about 175 kton is projected to increase to 288 kton by 1996. Owing to their good performance and recyclability, the consumption of TPO increases by 12% annually. The materials are used by the automotive industry (84% of TPOs are used in bumpers, dashboards, fascias, tubing, etc.), in consumer products, construction, geomembranes and medical applications [111]. The recently proposed method of fireproofing PP/PE TPO blends (by addition of tetrabomobisphenol-A-dibromopropyl ether) will further extend their usage [112].

TPOs are developed for a broad range of applications, ranging from flexible to rigid and tough. Most blends are prepared by mechanical mixing, with morphology dependent on the compounding and processing parameters [113–116], and, more recently, by direct sequential polymerization of propylene and ethylene–propylene mixtures [117]. However, as yet, these reactor–made blends (RTPO) do not have the same range of properties expected from the compounded TPOs (CTPOs) [118]. The popularity of TPOs originates from excellent low temperature impact strength, good stiffness, flowability, low shrinkage, wide range of modulus, low density, and good moisture, chemical, solvent and electrical resistance. In many cases, TPOs are reinforced by addition of glass fiber and/or mica flakes [119].

The behavior characteristic of thermoplastic elastomers is also shown by the olefinic polymers containing sulfonic moieties [120]. In PP blends with sulfonated EPDM, the presence of ionic groups significantly improved the solid state properties, such as tensile and tear strength [121]. The dynamic mechanical tests showed that only the glass transition temperature was affected by blending; T_g(EPDM) decreased from −45 to −55 °C. SEM observations demonstrated that for compositions ranging from 25 to 75%, both phases were continuous [122].

Blends of PP with aPP were studied by Romankevich and Frenkel [123]. Upon addition of aPP, the T_m of PP decreased systematically (see Figure 3.1). A plot of the free energy of mixing versus ϕ computed from these data showed a symmetric NDB, evidence of blend miscibility.

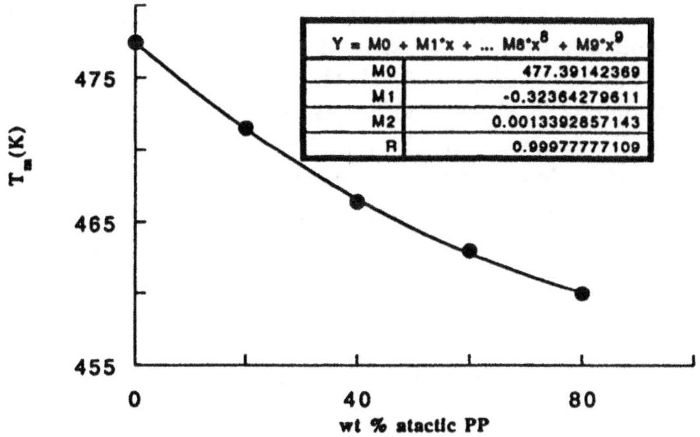

Figure 3.1 Melting point of isotactic PP in its miscible blends with atactic PP. The solid line represents the second order polynomial with numerical values of parameters listed in the insert; data from [123].

3.3.2 Melt flow

It is impossible to generalize the flow behavior of PP/elastomer blends. Blending different types of elastomers makes any such attempt impractical. However, as for other blends, the flow curves are expected to show a pseudoplastic behavior with yield stress at higher elastomer loading. Since the rubber phase viscosity is usually higher, unless the blends are prepared in extensional field, the phase inversion should occur at relatively high elastomer concentration. The η versus ϕ dependence is expected to be controlled by the emulsion-like and interlayer slip flow mechanisms [11].

It was shown that in PP/EPR blends, the minimum in melt density and capillary viscosity occur at the same ϕ [124]. The final morphology depended more on processing than on the composition. The capillary flow data of these blends indicated pseudoplastic flow without apparent yield stress. The constant stress viscosity versus composition plot showed a complex dependence, engendered by presence of both the emulsion-like and the interlayer slip effects [125].

The melt flow, morphology and mechanical properties of PP/HDPE/EPDM blends were discussed by Chaudhary, Varma and Varma [126]. Melt rheology of an electron beam irradiated PP/EPDM system was studied by Van Gisbergen, Hoeben and Meijer [127]. The high frequency viscosity of irradiated blends was found to be lower than that of the original mixtures, but at low frequency the opposite as true. Aggregation of the crosslinked EPDM domains was given as an explanation for this behavior.

3.3.3 Morphology

The multiphase character of PP/EPDM blends was confirmed by both dynamic mechanical tests and microscopy by Karger-Kocsis and Kiss [128] and later by Pukánszky *et al.* [129, 130]. The degree of dispersion was found to be nearly as high as that observed for PP block copolymers in which the ethylene-rich blocks separate out. The particle size of EPDM, as determined in SEM, was found to depend on the viscosity ratio. For $\lambda \leqslant 2.7$, the average diameter was $d \approx 0.5$–$0.6\ \mu m$, increasing abruptly for higher viscosity ratios [131]. The particle size distribution was influenced by the chemical composition of EPDM. A skin–core morphology was observed in injection molding of these blends [132].

Fortelny *et al.* [133] examined the structure of PP/EPDM and PE/EPDM blends in the full range of composition. In parallel, they measured the torque required to mix the components in a small mixing chamber. The concentration dependence of the torque showed a negative deviation, except for $w > 80$ wt% EPDM where a positive deviation was observed. Micrographs of the freeze-fractured surfaces showed the following changes of morphology with EPDM content: for w(EPDM)$\leqslant 20$ wt%, small EPDM spheres were observed; for $w \approx 40$ wt%, large and deformed EPDM particles; for $50 < w < 70$ wt%, a co-continuous structure. The specific volume of the blends was found to be significantly larger than what could be predicted from the additivity rule. The results for both PP and PE systems showed that blends with a lower viscosity EPDM showed phase co-continuity in a wider range of concentration than those with high viscosity ones. The morphology of solution cast PP blends with either random or sequential EPR was reported to be similar [134]. The complex morphology and its effect on PP/EPDM performance was discussed recently by Hoppner and Wendorff [135] and by Pukánszky *et al.* [129, 130, 136].

Bartczak *et al.* [137] investigated the nucleation in PP/EPR blends in OM. During melt blending there was migration of the heterogeneous nucleating agent from the EPR to the PP phase. The self-seeded nucleation was observed to decrease during blending. Recently, the influence of the EPR molecular structure on the morphology of its blends with PP was studied by D'Orazio *et al.* [138, 139]. Several PP/elastomer blends in a wide range of composition were cast from solution [140]. The smallest particles, $d \approx 40$–80 nm, were observed for PP/EPR, larger, $d \approx 50$–150 nm, for PP/polybutadiene (PB), and the largest for PP/styrene–butadiene rubber (SBR) blends. Thus, in PP/EPR blends, the dispersion was found to be comparable to the calculated molecular coil size. In the remaining blends, a three-phase structure was observed: rubber, compatibilized layer and PP.

The morphology and performance of PP/PB blends were discussed recently by Gupta and Ratnam [141] and that of PP/PB-1 by Wenig and Schoeller [142] and Siegmann [143]. The latter studied the crystallization and performance of PP/PB-1 blends and showed, using DSC, that crystallization of both polymers was affected by the other's presence. Two distinct effects of PP on the PB-1 crystallization were noted. At low PP content, PP spherulites acted as the nucleating agents of PB-1, increasing its crystallization temperature, T_c. At high PP content, the PP amorphous phase played the role of a polymeric diluent, reducing T_c(PB-1). The dynamic mechanical tests suggested that the system is immiscible, although a small shift of the melting points was observed. On the other hand, a single glass transition peak, T_g, was taken as evidence of miscibility in the amorphous region. More recently, Lee and Chen [144] studied similar blends. The authors observed enhancement of crystallinity of PB-1 resulting from nucleation by PP spherulites.

Martuscelli, Silvestre and Abate [145] studied the morphology of films of PP mixtures with either EPR or PIB. Both elastomers acted as nucleating agents for PP. The presence of EPR caused an increase of T_m(PP), explained by PP-defects extraction by the elastomer. PIB slowed the growth rate of PP spherulites, suggesting some miscibility between the two polymers. The micro-morphology and the mechanical properties of PP blended with *trans*-polyoctanylene were investigated by Wenig and Fiedel [146], while those of PP blends with hydrogenated oligocyclopentadiene were studied by Martuscelli *et al.* [147], Marcandalli *et al.* [148], and Cimmino *et al.* [149].

3.3.4 Mechanical properties

As indicated above, there are numerous publications dealing with relationships between morphology and performance of PP/elastomer blends

[25, 104, 105, 150, 151]. The optimum PP/EPR composition depends on what is of particular importance for a given application (e.g. impact strength, modulus, strain at break, etc.) as well as on the component molecular parameters. However, addition of $w(\text{EPR}) \approx 20$ to 30 wt% usually leads to good balance of properties [116, 125, 140].

The effect of MW and composition on the tensile properties of PP/EPDM blends was examined by preparing binary blends using three PP and two EPDM resins of different MW and measuring their properties over the entire composition range [152]. Positive deviation was observed for the tensile modulus. The optimum composition range varied from 5 to 20 wt% EPDM, depending on its type and MW of PP. The tensile results were found to follow Kerner's equation. The morphological examination suggested that PP was the matrix up to 75 wt% of EPDM.

Using two-dimensional finite element analysis, Kikuchi *et al.* [153] studied the origin of rubber elasticity in a PP/EPDM system. The key for strain recovery was found to be that during deformation the ligament matrix between rubbery particles does not exceed the elastic limit.

Kolarik *et al.* examined the dynamic mechanical properties of binary and ternary blends of PP, PE and EPDM [154, 155]. The dynamic moduli versus temperature curves of PP/EPDM blends showed that the transition temperature, characteristic for each polymer, remained unaffected by blending (indicating system immiscibility). A comparison of the experimental shear and tensile moduli with the theoretical predictions suggested that phase inversion occurred at $\phi(\text{EPDM}) \approx 0.5$, while PP preserves certain phase continuity up to $\phi(\text{EPDM}) \approx 0.8$. Within this higher range of concentrations, $0.5 \leqslant \phi(\text{EPDM}) \leqslant 0.8$, higher than predicted moduli values were recorded. Examining the ternary blends, the authors observed that the minor polymeric component was occluded within EPDM particles; 20 vol% of EPDM could accommodate 20 to 30 vol% of either PE or PP. The effect of addition of the minor component on the mechanical properties was similar to the addition of rubber. Recently, Zhu *et al.* [156] studied microcrystallinity of PP/EPDM blends and its effects on the mechanical performance of automobile bumpers and dashboards.

PP/EPR blends were studied by Lee and Lu [125], Coppola *et al.* [157] and Ramsteiner [158]. Lee and Lu, after adding 30 wt% EPR to PP, observed an increase of the impact strength by a factor of 26, accompanied by a decrease in Young's modulus, tensile strength and heat deflection temperature [125]. Coppola *et al.* studied morphology, mechanical properties and drawing behavior. Addition of $w(\text{EPR}) \geqslant 10$ wt% was reported to decrease PP spherulite size [157]. The tensile stress–strain properties were measured for different compositions, T_c, and drawing temperatures, T_{dr}. The curves of the modulus versus T_{dr} showed a break, indicating a transition in the drawing mechanism. This transition temperature was almost independent of EPR content, but increased with T_c.

Recently, the effect of morphology on stress whitening and impact strength at temperatures below T_g was studied by Ramsteiner [158]. The author showed that the mechanism for improvement of the impact strength is related to crazing around the EPR particles. The whitening occurs since crazing and subsequent formation of cavities at high strains scatters light. It was demonstrated that the size of the dispersed phase has a significant effect on both the impact strength and strain whitening. The effect of annealing on the micro-morphology of PP/EPR blends, as well as on the impact strength and the dynamic mechanical properties, were reported by Ito, Mitani and Mizutani [159]. It was observed that annealing increased T_g, crystallinity, yield and tensile strength, but decreased the strain at break, brittleness temperature and impact strength. The effect depended on molecular parameters of the system as well as on annealing conditions. Varga and Garzó [160] reported on the mechanical properties of PP/EPR blends in which the PP existed in its β-crystalline form.

Brittle fracture mechanics studies on PP/EPR, with and without HDPE, were reported in [161–163]. The mechanism of fracture was found to be mixed, crazing and shear yielding in different proportions, depending on specimen preparation (annealing) and test conditions [164]. The apparent toughness was found to increase with rubber content, while the calculated plane stress toughness remained constant. The thermal treatment affecting the spherulite size had a pronounced effect on toughness.

Performance of a three-component, immiscible system is complicated by the presence of three interfaces requiring simultaneous optimization of three different interfacial tension coefficients. In many cases development of well performing polymer blend, then addition of a filler or a reinforcing material leads to disaster. Since addition of solids usually changes the location of compatibilizers, the three interfacial coefficients must be optimized at the same time. Thus, performance of reinforced PP blends is difficult to predict and control. Gupta, Srinivasa and Kumar [165] and Theisen, Hess and Kosfeld [166] discuss some of these difficulties.

3.4 BLENDS WITH POLYAMIDES

3.4.1 Introduction

Polypropylene/polyamide blends belong to a large and growing family of polyolefins (PO) alloyed with PA. In a sense, toughened PAs (e.g. a blend of PA with $w = 2.5$ to 5 wt% of PO or an ionomer) are precursors of PO/PA alloys, where usually about 40 wt% of PO is dispersed in a PA matrix [167–170]. There are several PA/PO blends available on the market. Since 1985, PP has been used as PO with increasing frequency and PA-6 as PA [2]. Other PA/PO blends, for example, PP with PA-66, PA-12, or with aromatic polyamides are also available. There are five

PP/PA manufacturers in Europe, nine in Japan and one in the USA. The reasons for blending PO and PA are to bridge the property gap between the two resins, to reduce the cost, to decrease the water absorption of PA, to improve processability of PP, and to benefit from the thermal and mechanical properties of polyamide while improving its impact resistance. Thus, the often cited advantages of PA/PP blends are: low moisture absorption, improved processability, good impact resistance, and flexural modulus [171–174].

In most cases, alloying is a two-step reactive process, involving modification of PO, followed by mixing with PA. The simplest, but not necessarily the best, method is maleation of PO in the presence of peroxides. Other acids or anhydrides (e.g. fumaric, acrylic, methacrylic) have also been used [173–178]. A new commercial compatibilizer based on ethylene–acrylic ester–maleic anhydride–glycidyl methacrylate was recently described by Costin and Billet [179]. The compatibilizer may also be formed by reacting the acidic copolymer with an oligopolyamide [171]. The compatibilization of PA/PP blends by addition of the first stage acidic copolymer (e.g. maleated PP-wax, maleated styrene–butadiene–styrene or maleated styrene–ethylene/butylene–styrene tri-block copolymer) has also been carried out. A new twin screw extruder with several test points for sampling the reacting mixture was recently described by Nishio *et al.* [178]. As an example, the authors considered morphology development in the PP/PA-6/m-PP system as a function of MW (PA-6) and location along the screw.

Owing to the adverse effect of compatibilization on blend crystallinity, its extent must be optimized, then controlled. This usually results in blends whose morphology depends on the processing and thermomechanical treatment. This, on the one hand, allows for development of the desired final product morphology, and thus optimization of the material performance, and on the other, for possibilities of shear coalescence and separation of the ingredient on a macro scale. However, if higher stability of morphology is required, either a higher level of compatibilization or a decrease of PP content may provide a solution. In this case the property versus profitability relation would also change [20, 180].

3.4.2 Melt rheology

Several reports on the melt flow behavior of PA/PP have been published [19, 175, 180–187]. The shear flow data frequently indicate the presence of an apparent yield stress, σ_y. For PP/PA blends, the concentration dependence of η at constant shear stress, σ_{12}, usually shows NDB, indicating the predominant effect of the interlayer slip [19, 175, 180]. The slip (and NDB) increases with σ_{12} [183–188]. On the other hand, the η versus ϕ dependence for compatibilized alloys at low σ_{12} may show PDB, NDB at high σ_{12}, and PNDB at the intermediate stress levels. In the case of compatibilized blends, interpretation of the η versus ϕ dependence may be complicated by the chemical reactions occurring between PA and the compatibilizer. Furthermore, the compatibilizer alone may be highly viscous, causing rigidification of the interphase and, as a result, a suspension-like behavior of the alloy leading to PDB. Evaluating the rheology of compatibilized blends, one must consider: (i) variability of the morphology with imposed stress, (ii) flow-dependent kinetics of phase ripening, and (iii) effect of the third phase (interphase) with its own rheological characteristics.

Compatibilization of PA/PP blends is achieved when a relatively large quantity, $w_3 \geqslant 4$ wt%, of the first stage (acidic) compatibilizing agent is used [175, 189, 190]. Since such a compatibilizer reacts with PA, the resulting concentration of PA–PP copolymer found in the blends varies from about 10 to 20 wt%. At this level, owing to its relatively high molecular weight, the copolymer may significantly increase the system viscosity [192]. On the other hand, it has been observed that since the level of molecular entanglement in the interphase is low, the interphase viscosity is much lower than that of the two polymers [193, 194]. Thus, the degree of rheological change caused by the compatibilization depends on the compatibilizer, compatibilization method and its extent. For example, Scholz, Froelich and Muller [195] reported that in their PP/PA-6 blends the dynamic shear response was independent of the level of compatibilizer (PA–g–PP).

Since compatibilization changes the interphasial properties, and as a result, modifies the interaction between the dispersed phase drops, an effect on σ_y may be expected. Indeed, for compatibilized PA/PO blends, σ_y was quite apparent [196, 197].

Hayashida and Yoshida [188] used an extruder type capillary rheometer to measure the flow properties of PP/PA-6 blends over the whole concentration range. Strong NDB in the η versus ϕ dependence was observed at all shear rates. The lowest viscosity was obtained at 25 wt% PA-6.

Microscopic examination made it possible to postulate that here the surface contact between the two phases and, consequently, the interlayer slip, were the largest. Similar blends were studied by Liang *et al.* [183]. The viscosity composition plots showed a PNDB dependence. The first normal stress difference for the blends was higher than that of the neat resins.

Within the region of phase inversion the extrudate swell ratio, B, plotted as a function of composition, was found to go through a local maximum [184–186]. It is worth recalling that, with the exception of dispersions with solid-like particles, similar behavior has been observed for most PABs. As a rule, B in blends is large and shows PDB. The effect is not related to deformation of the macromolecular coil (as it happens for homopolymer melts), but rather to the form recovery of the dispersed phase (strain recovery or 'extrudate shrinkage') [2].

A summary of round-robin tests carried out by the Versailles Project on Advanced Materials and Standards, Technical Working Party on Polymer

Blends (VAMAS TWP-PB) presents extensional, dynamic and steady state shear viscosity data for a PP/PA-6 commercial alloy [198]. The report concludes that: (i) the melt flow properties depend on the extent of sample drying; (ii) there is a lack of agreement between steady state, dynamic, and extensional viscosity data; (iii) there is a lack of agreement between the first normal stress difference and the storage modulus; and (iv) the elastic responses in steady state flow are much larger than those observed for single phase systems. All these observations can be explained by the two-phase nature of the alloy. Despite the complex viscoelastic nature of the material the dynamic test results reported by the six different laboratories were quite reproducible.

Recently Utracki and Sammut [180] reported on the melt flow behavior of PP/PA-6 alloy in dynamic and steady state shear flow as well as in uniaxial extension. It was observed that: (i) the blend required nearly ten times longer drying than PA-6; (ii) owing to polycondensation during the dynamic tests, log G' and log G'' increased linearly with time; (iii) the dynamic viscosity, η', showed linear viscoelastic behavior, without yield stress; (iv) in capillary flow at $T=225$ to $250\,°C$, η was found to be insensitive to temperature (leading to poor agreement with η' or the complex viscosity, η^*); the poor correlation was a consequence of the flow modification of blend morphology caused by large strains encountered during the capillary flow; (v) owing to strain hardening in uniaxial extension, an agreement between the dynamic data, $3\eta'$, and the equilibrium value of elongational viscosity, η_E, measured in uniform constant rate of deformation field can be found only within the range of low deformation rates, well below those experimentally accessible; (vi) the directly measured η_E was one order of magnitude smaller than that calculated from the entrance pressure drop correction by means of the Cogswell relation; the disparity is related to differences in blend morphology between the two types of flow. Thus, it was concluded that the capillary flow measurements are not suitable for characterization of PABs and should not be used for this purpose. Furthermore, the Cogswell method is not applicable to multiphase liquids.

3.4.3 Morphology

Uncompatibilized PA–6/PP blends are coarse mixtures with large particles ($d \leqslant 50\,\mu m$), easily deformable into fibrils [187, 189, 192, 199, 200]. Near the phase inversion composition, the morphology is concentration, stress and heat sensitive. Upon compatibilization, the drop size is reduced to about $1\,\mu m$. In commercial alloy, even at $w(PP)=40\,wt\%$, PP ($\lambda > 1$) forms the dispersed phase with spherical drops of the number average diameter, $d_n = 1.1 \pm 0.6\,\mu m$ [191, 201]. However, in spite of the compatibilization, processing of these well-dispersed blends was found to lead to flow segregation, coarsening of morphology and fibrillation [20, 180]. Micrographs of the PP/PA-6 pellets showed spherical particles of PP embedded in the PA matrix, while in extruded sheets, PP domains were large, extended and oriented. The degree of PP phase fibrillation was found to depend on the depth: near the surface the fibers seemed fused, entrapping fibrillated PA-6 phase, while at the central plane, large PP drops were only slightly deformed, with the axial ratio of the particles rarely exceeding two.

The phase morphology and melt rheology of PP/PA-6 blends have been discussed in terms of their impact on the structure development during fiber spinning [199]. At higher temperatures, phase coarsening was observed, specially at $w(\mathrm{PP}) = 25$ to 50 wt%. Reduction of the drop size with increasing shear rate during capillary flow was also reported. The molecular orientation in each phase was determined by wide angle X-ray scattering. The results showed that orientation in PP was independent of composition while the reverse was true for PA-6. Romankevich *et al.* [200] and Yakovlev, Shevchenko and Romankevich [185, 186] used spinnerets with jet stretch producing ultrathin fibers, independent of the size of the spinneret hole. Under these conditions, the PP dispersed phase had fibrillar structure as opposed to spherical particles obtained without jet stretch. On the other hand, reducing the spinneret holes or increasing the spin rate also caused the spherical particles to deform into ellipsoidal particles without jet stretch. Fibrillar structures were also seen for a spinneret hole 0.25 mm diameter.

Addition of an ionomer (a terpolymer of ethylene, neutralized methacrylic acid and isobutyl acrylate) to blends of PP with PA-6 significantly reduced the dispersed phase size and improved mechanical properties [189, 190]. In blends containing $w(\mathrm{PA\text{-}6}) = 10$ wt%, the maximum reduction of particle size was observed for 0.5 wt% of the ionomer. The final degree of dispersion depended on the mode of ionomer addition: one step (mixing of all ingredients at once) or two steps (addition of ionomer to PA-6 then incorporating PP). The two-step method gave smaller drops when PP was the minor phase but larger ones when PA-6 was the minor phase.

Several studies of melting in PP/PA-6 blends have been published recently [202–204]. T_m(PA-6) in the blends was found to be depressed by pressure. In another study, addition of PP to PA was observed to decrease T_m by different amounts; by 4 °C in PP/PA-6 blends and by 6 °C in PP/PA-6/(m-PP) blends [187]. In PA-6/(m-PP) blends, increasing PA-6 content caused a decrease of T_m(PA-6), but an increase of that of m-PP. The authors speculated that this was due to dissolution during compounding of the grafted PP molecules in PA-6 phase. A decrease of total crystallinity upon addition of a compatibilizer, as well as with compounding time, was also reported [20].

3.4.4 Mechanical properties

The tensile properties of PP/PA-6/ionomer have been studied as a function of composition, method of blend preparation and draw ratio, DR [189, 192].

The modulus as well as the stress and strain at yield for blends containing w(PA-6) = 10 and 30 wt% was reported to go through a maximum at DR ≈ 2–3, while for the compatibilized blends, the same properties showed a steady increase up to DR = 7. For the blend PP/PA-6/(m-PP), Sato *et al.* [205] reported that the tensile strength and modulus increased monotonically with increasing PA-6 content while the tensile strain at break showed a large negative deviation. The effect of PA-6 addition on the morphology and performance of PP fibers was studied by Grof, Sain and Durcova [206]. While at very low concentration of PA-6, a good fiber was obtained, further increase was found to severely lower the melt strength, probably due to lowering of blend viscosity. Addition of maleated PP to the blend significantly improved drawability and fiber performance.

Six laboratories participated in a round-robin testing on dynamic mechanical behavior of a PP/PA-6 commercial alloy under the auspices of the VAMAS TWP-PB. The specimens were prepared by extrusion, injection or compression molding. The greatest difficulty in comparing the data arose from variation of the moisture content. In the extruded specimens a minor effect of flow orientation was detected. Overall, there was good agreement between the contributing laboratories for the frequency dependence of the transition peaks, but the level of agreement between the absolute values of the moduli was poor. This discrepancy is suspected to be partially due to the differences in sample geometry and in the processing conditions [207]. The dynamic spectra were studied at $T = -150$ to $+150$ °C and frequency $\nu = 1$ to 90 Hz [180]. Three transitions were observed, one corresponding to T_g(PP) = 3 ± 1 °C, and two others associated with vibrations of PA-6 chains. Their initial values, $T_1 = -68$ °C and $T_2 = 26$ °C, upon drying increased by 14 and 30 °C, respectively; T_1 originated from the water–amide vibrations, whereas $T_2 = T_g$(PA-6). An increase of T_g(PA-6) by 3 °C was observed, indicating that both PP and the compatibilizer had little, if any, effect on the amorphous phase. Owing to lower crystallinity in the PP/PA-6 blends (than that in the neat resins) their moduli were depressed. The concentration dependence of dynamic moduli was interpreted using Equation 9. Contrary to expectations, it was found that the equation constants, α and δ, are temperature-sensitive. Their variability was most likely caused by changes in mobility of the interphasial layer.

The dynamic mechanical properties of PP/PA-11 blend were investigated by Liang and Williams [208]. It was found that the dynamic response of blends prepared with PA-11 and with its acrylic acid functionalized homologues were quite different. In a wide temperature range the storage modulus of the latter alloy was a linear function of concentration, whereas that of the former showed a strong negative deviation. The data did not show any variation of T_g with composition, but the sub-T_g, β-transition provided evidence that compatibilization primarily affected the PA-11 phase.

Purely brittle fracture was observed in Charpy impact strength tests on 2-mm-thick PP/PA-6 specimens [180]. The critical strain energy release $G_c = 7.9\,\text{kJ/m}^2$ was calculated following Plati and Williams' postulate that the energy observed at fracture, U, is a linear function of BDF (where B is specimen thickness, D its depth and F a tabulated parameter) [23, 24]. The value is comparable to that of neat PA-66 resin ($G_c = 8.2\,\text{kJ/m2}$), but significantly lower than that for supertough PA blends ($G_c = 75.7\,\text{kJ/m2}$) [209].

3.5 BLENDS WITH POLYCARBONATE

3.5.1 Introduction

Blending of PP with PC is at the stage where PP/PA blends were in the mid 1980s. Several industrial laboratories in Europe, Japan and the US are actively looking for compatibilization methods that will make it possible to explore blend performance within the whole composition range. Addition of about 5 wt% of PP to PC was found to reduce melt viscosity, thus improving processability (e.g. for injection molding), as well as increasing the impact strength (especially in thick sections) and modulus, reducing notch sensitivity and mold shrinkage [210]. On the other hand, addition of about 10 wt% of PC to PP improved its processability and product appearance, as well as increased hardness, modulus, and impact strength, reduced notch sensitivity and mold shrinkage [210, 211]. The key for achieving all the cited advantages is control of PC concentration, its particle size and degree of deformation (particle aspect ratio). As expected, the strain at break versus ϕ for these two component blends showed strong NDB [210].

There are already several patents describing PC compositions with $w(\text{PP}) \geqslant 10$ wt%. GE Plastics [212, 213] reported that addition of an acrylic polymer enabled an increase of PP content that led to better gasoline resistance. Addition of styrene–(ethylene/butylene)–styrene triblock copolymer (SEBS) allowed the use of up to 40 wt% of PC (optimum 15%) in blends for packaging film applications [214]. In [215] blends containing 20 to 80 wt% of either PP or PC compatibilized by a mixture of ethylene vinyl alcohol (EVAl) and SEBS grafted with either maleic anhydride or fumaric acid are described. Excellent mechanical properties, impact strength, heat distortion temperature, moldability, and solvent/chemical resistance have been claimed for blends containing 5 to 95 wt% of either PP or PC compatibilized with different compatibilizing systems, including SEBS, styrene-acrylonitrile grafted EPDM, m-PP, polystyrene containing oxazoline moieties, polybutyleneterephthalate, etc. [216]. A preparation of PO–PC–PO tri-block copolymer, to be used as compatibilizer in PO/PC blends, has been patented [217]. Good impact properties of cycloalkylenebisphenolpolycarbonate blends with PP were also claimed recently [218]. The blends also showed good high temperature performance.

Recently Liang and Williams [219] studied PC/PP blends containing 4% graft or block copolymers: poly(propylene-g-styrene), poly(propylene-g-styrene/acrylonitrile), poly(propylene-g-styrene/n-butylmethacrylate), poly(propylene-b-*n*-butylmethacrylate), and SEBS. The effects of these compatibilizers on the crystallization, relaxation behavior and mechanical properties were reported. Significant improvement of blend performance was observed.

3.5.2 Melt rheology

Information on the melt flow behavior of PP/PC blends is rather scarce. By contrast, owing to VAMAS TWP-PB activities, the viscoelastic properties of PC/LLDPE blends have been extensively studied in the molten [220–222] as well as in the solid state [213–225]. As the data of Liang and Williams [219] demonstrate, the latter system can be used as a model for PP/PC mixtures. As seen for PP/PA blends, the addition of compatibilizer does not necessarily eliminate the interlayer slip.

3.5.3 Morphology

There is no doubt that in antagonistically immiscible blends such as PP/PC, control of the morphology determines the blend performance [211]. Thus, there are several publications dealing with preparation and characterization of morphology in PO/PC blends [213, 216]. Addition of PC to PP leads to enhancement of PP-crystallinity and a small increase in T_c. The effects were significantly more pronounced upon addition of a compatibilizer [219]. The compatibilization was found to be limited to the amorphous phase.

The size and size distribution in PP/PC blends were studied in the full range of composition by Favis and Chalifoux [227]. The authors reported that the degree of dispersion depended mainly on λ, particularly when the high viscosity polymer formed the matrix. Both the particle size (ranging from about 1 to 13 μm) and the polydispersity increased with the minor phase content. The phase inversion region was affected by λ as well as by the relative elasticity of the components. In the later studies the effect of processing on morphology was investigated [228]. Large differences in morphology were reported between the center and the outside region of the extruded strand, indicating the dominant role of the die zone for structure formation. Elongated fibers were present at the rim while mostly droplets were seen in the center.

3.5.4 Mechanical properties

Dynamic mechanical as well as tensile properties of PP/PC blends were measured by Liang and Williams [219]. The relaxation behavior of the blend was found to be additive, calculable from those of the two major components, with compatibilizer playing only a small role. The compositional dependence of the heat deflection temperature, modulus, yield and fracture strength for blends containing $w(\text{PC}) \geqslant 25$ wt% all showed a negative deviation from the additivity rule. Addition of a compatibilizer clearly improved the blend performance. The best overall effects were observed with poly(propylene-g-styrene/acrylonitrile).

Fisa, Favis and Bourgeois [229] investigated the effect of processing on morphology and mechanical properties in PP blends with $w(\text{PC}) = 0$ to 40 wt%. In spite of poor interphase adhesion, enhancement of the modulus and strength was observed; the modulus followed Equation 9. Upon addition of PC, the maximum strain at break as well as the weld-line strength in injection molded specimens sharply decreased.

3.6 BLENDS WITH POLYESTERS

Under this title, blends with thermoplastic polyesters (PESTs) as well as liquid crystal polyesters (LCPs) will be discussed. Small amounts of PP have been used for toughening PESTs [230, 231]. Blends of PP with poly(cyclohexylenedimethylene-co-ethyleneterephthalate) (PCTG) were reported to have good impact properties at $w(\text{PP}) = 30$ and 50 wt% [232]. Blends of PEST (40 to 50 wt%) with PO (40 to 50 wt%) and SEBS (10 to 20 wt%) are claimed to show a good balance of properties, making them suitable for manufacturing tubing, hoses, wire and cable jacketing, etc. [233].

At present, most PP/PEST blends are formulated within the low concentration region of dispersed phase: PP in PEST and vice versa, where the mutually beneficial effects are evident [234–236]. However, in spite of the early patents and the present activities, there is no commercial PP/PEST blend on the market. As in the case of PP/PC blends, the problem rests in finding a good compatibilizer, allowing for exploration of the mid-range compositions of PP/PEST mixtures. The current use of m-PP does not engender the desired performance.

The PP/LCP blends are relatively new. Their usefulness is based on the uniaxial or biaxial orientation of LCP crystals dispersed in PP matrix. Thus, exploitation of these materials is not conditional on the chemical modification of the interphase, but rather on development of new, advanced methods of processing [237–240].

Flow curves for PP with polyethyleneterephthalate (PET) were measured in a single screw extruder [234]. Two compositions, $w(\text{PET}) = 50$ and

70 wt%, showed lower viscosity and smaller entrance pressure correction than those for PP. Although PET had lower η than the blends, they extruded at higher volumetric rates, indicating that the output was primarily controlled by melting. It was shown that even if the sheath-and-core monofilaments of PP/PET blends could not be oriented (because of poor adhesion between the two polymers), melt blending still produced useful oriented product [234]. The maximum draw ratio for PET was 7, whereas it was 11 for PP/PET blends. The dynamic mechanical testing of blend monofilaments, containing $w(\text{PET}) = 50$ and 70 wt%, showed that $T_g(\text{PET})$ remained constant and the dynamic moduli was approximatively additive.

Flow properties of PP/LCP blend in the whole composition range were determined by Ye *et al.* [236]. At concentrations $w(\text{LCP}) \geqslant 1$ wt% an apparent yield stress was observed. Plots of constant stress η versus composition indicated significant interlayer slip with $\beta > 1$ (see Equation 7). The influence of processing on morphology and properties of immiscible PP/LCP blends was studied recently [236–240]. These materials undergo two-step (domain and molecular) orientation. Both are required for good performance.

Addition of LCP to PP followed by two-step orientation was found to increase the tensile modulus in the machine direction by one order of magnitude, and decrease its value in the transverse direction [239]. On the other hand, without molecular orientation, no significant improvement was obtained; and the maximum strain at break decreased with LCP loading [236]. Compatibilization of PP/LCP blends with m-PP significantly improved these properties [240]. The plot of tensile modulus versus composition showed either negative or positive deviation from additivity; values comparable to those of PP/glass fiber systems were obtained.

3.7 BLENDS WITH STYRENICS

3.7.1 Introduction

Blends of PP with polystyrene (PS) are primarily of academic interest [241, 242]. They are nearly ideal for studying the effects of numerous variables on the blend morphology. The practical use of two-component PP/PS blends is based on flow segregation causing migration of the PP phase to the surface, thus allowing for higher temperature applications and grease resistance [243]. Blends of PP with high impact polystyrene (HIPS) and styrene–butadiene–styrene tri-block copolymer (SBS) show good mechanical properties and large strain at break [244–245]. The three-component blend of styrene-butadiene block copolymer (SB) with PP and EPDM was found to have an exceptionally good balance of properties, particularly distortion and solvent resistance, as well as high temperature retention of hardness [246]. Blending of LLDPE with PP and PS was also found to be advantageous

[247]. PP blended with poly(styrene-co-terpene) is useful for production of biaxially oriented, heat sealable packaging film [248].

3.7.2 Melt rheology

Formation of PP/PS blend structure was studied by Miroshnikov and his collaborators [249–251]. Blends with $\lambda=0.11$, 0.17, 0.55 and 4.10 and $N_1(\mathrm{PS})>N_1(\mathrm{PP})$ were prepared using four PP resins with one PS. Strong interlayer slip effects were demonstrated. Following on these works, Cohen and Schroeder attempted to explain the viscosity anomalies in PP/PS blends with morphology [252]. They extruded through capillaries pre-mixed and layered PP/PS blends, observing formation of concentric axial layers and reduction of η. The drop in viscosity was proportional to the number of layers.

The capillary flow of PP/PS binary blend was studied by Nishimura [63]. The plot of log η versus ϕ showed NDB at all σ_{12} levels. By contrast, the entrance pressure drop showed PDB. The influence of mixing on the capillary flow seemed weak. The photomicrographs of extrudates showed an annular structure of the fractured surface, with the less viscous component primarily concentrated on the outside. Deri, Genillon and May [253] made similar observations: the authors observed NDB for the η versus ϕ plots and PDB for B versus ϕ at all shear rates.

Rheology of PP with styrene copolymers – SBS, SEBS, acrylonitrile–butadiene–styrene (ABS) and styrene–acrylonitryle copolymer (SAN) – have also been studied [249, 254–264]. The data indicate flow behavior characteristic of PABs. Addition of 10% PS, HDPE or SEBS reduced PP flow instabilities as well as extrudate distortion.

3.7.3 Morphology

Morphology of PP/PS blends containing $\phi(\mathrm{PP})=0.15$ and 0.25 was studied by Miroshnikov and Williams [249]. Four systems differing in viscosity and elasticity ratios were extruded at 200 °C and $\sigma_{12}\approx 1$ to 100 kPa. Extended structures were obtained in all cases. The authors characterized the morphology by citing percentages of specimens in which fibers, ribbons, layers, branched, reticulate and occluded structures were present. Van Oene elasticity difference theory [262] was used for interpretation of the results. The authors noted that short capillaries were 'more efficient for forming pure fibrous structures'. However, they did not associate this phenomenon with the extensional forces at the capillary entrance.

The stability of the blend structure was also discussed. Decomposition of fine fibers into rows of monodispersed drops was documented during the 1930s by Taylor [263, 264], who postulated that the capillarity instability mechanism is responsible for the breakup. Morphology of extruded PP blends with 5 wt% poly(styrene-co-2,2,6,6-tetramethyl-4-piperidinyl methacrylate) (PDS) was studied by Fan and Hou [265]. As in the works cited earlier, the PDS drop size, deformation, and drop diameter were found to depend on σ_{12}, capillary length-to-diameter ratio, and MW of PDS.

The effect of PS on the crystallization of PP was investigated by Verbraak, Van Dam and Van der Vegt [266] and by Wenig, Fiedel and Scholl [267]. It was found that PP crystallization is affected by the presence of PS. Upon increase of concentration, the nucleation density of PP decreased while the spherulitic growth rate (as well as the position of thermal peaks) remained constant. The atactic PS constituted the external matrix that facilitated the processing and allowed transfer of stresses to the PP. This resulted in formation of highly oriented PP fibers with modulus ten times the value of the unoriented sample.

Owing to miscibility of PP with EPR (containing $w \ll 10$ wt% of ethylene) and with PB-1 it is expected that some SBR resins will be good impact modifiers for PP. However, as for blends with EPR and EPDM resins, here also addition of SBR is expected to affect the morphology of the system. The effects of SBR particle size, processing method, thermal treatment, etc., on micro- and macro-morphology were studied by Xu, Xia and Li [268]. Similar studies on PP blended with styrene–isoprene block copolymer were discussed by Hlavata *et al.* [269].

An extensive study on morphology development during compounding in a twin screw extruder has been undertaken by Utracki and his collaborators [10, 14, 270]. The study involves, on the one hand, development of a predictive computer program (based on the micro-rheological approach), and on the other a detailed experimental study of the variation of shape, size and polydispersity of the dispersed phase during the mixing. The model, within limitations imposed by the initial assumptions, well describes the process.

3.7.4 Mechanical properties

Owing to weak adhesion between PP and PS phases, the ultimate mechanical properties of the blends are poor. Blends of PP with acrylonitrile–butadiene–styrene (ABS) in the full range of composition were investigated by Markin and Williams [271]. The blend density was found to be lower than calculated from the component densities. Despite this, the modulus versus composition plot was sigmoidal, indicating good PP toughening upon addition of $w(\text{ABS}) \leqslant 30$ wt%. The impact strength values were poor. Interestingly, the unnotched specimens showed better impact strength than the notched ones. The difference was traced to shear fractionation of blend composition leading to virtual coating of the injection molded specimens by the low viscosity ABS component.

3.8 OTHER BLENDS

3.8.1 Polyurethane blends

Addition of polyurethane (PU) to PP was expected to improve its impact properties. To test this hypothesis several PUs were blended with PP at compositions: $w(\text{PU}) \leqslant 50$ wt% [272–274]. Melt flow, thermal, dielectric, and mechanical properties, density, morphological, X-ray and dynamic mechanical analyses have been performed. Good mechanical behavior was obtained for $w(\text{PU}) < 15$ wt%.

3.8.2. Nitrile rubber blends

By incorporating 2 wt% dimethylol-p-octylphenol into PP, then blending with nitrite rubber significant improvement in the tensile strength, tensile strength at break, and maximum strain at break were achieved [275].

3.8.3. Ionomers

Addition to PP of 1.5 to 20 wt% neutralized m-PP was found to result in material having good thermal stability, improved melt strength (particularly advantageous in thermoforming), and mechanical properties [276]. Blends of ionomers with PP-copolymers have also been studied [277].

3.8.4 Polyetherimide

Blends of PP with polyetherimide (PEI) were investigated in the full range of composition. Addition of $w(\text{PP}) = 10 \pm 5$ wt% significantly improved the notched Izod impact strength of PEI as well as its ultimate strength in flexural tests. At the same time decreases in unnotched impact strength, modulus and the tensile properties were recorded [278].

3.8.5 Polycaprolactone

PP blends with polycaprolactone (PCL) were examined by DSC, dynamic mechanical spectroscopy and OM [279]. Dynamic testing showed a two-phase structure with limited phase mixing even between the two amorphous phases. A small melting point depression of PCL was also reported.

3.8.6 Ethylene–vinylacetate copolymer

Crystallization of PP/ethylene–vinylacetate copolymer (EVAc) blends was studied by high pressure DSC [280]. A decrease T_m(PP) at low concentration of PP was reported. An upward shift of the PP crystallization peak upon addition of interfacial agent (m-PP) suggested that the latter mixed with the PP phase and nucleated its crystallization.

3.9 CONCLUSIONS

Polypropylene blends constitute one of the most rapidly growing segments of the plastics industry. Development of blends with polyolefins is the most advanced. However, even here there are many conflicting observations, e.g. regarding the effects of the molecular parameters on blend performance. On the other hand, PP alloys with engineering resins (e.g. polyamides, polycarbonate or polyesters) are at an early stage of development. There are several indications that the most popular method of compatibilization with maleated PP may not be ideal – there is intensive activity to find better compatibilizers and/or compatibilization methods.

The main source of difficulties in the development of PP blends is the 'antagonistic' immiscibility of this polymer with almost any other. Thus, compatibilization is required, which results in modification of the micro- and macro-morphology of the system. In many cases, addition of compatibilizer to the PP mixture with an engineering resin reduced the crystallinity, which in turn lowers the mechanical performance. Optimization of composition and blending method is necessary.

Like any other blend, the performance of PP blends is controlled by a multiplicity of factors. So far, optimization has been done on an empirical basis. There is an urgent need for development of a more fundamental approach to polymer blending. The new approach should permit optimization of the blend composition on the basis of the molecular considerations (chemical nature of the components, their molecular size and structure as well as composition). Furthermore, it should lead to selection of the best compounding method, considering the most appropriate morphology required by the blend performance. This can only be done by developing more coherent, international cooperation between 'compatible' research teams.

3.10 APPENDIX: SYMBOLS AND ABBREVIATIONS

ABS	Acrylonitrile–styrene copolymer
aPP	Atactic polypropylene
B	Extrudate swell ratio
CTPO	Compounded olefinic thermoplastic elastomer
d	Dispersed phase drop diameter
DR	Draw ratio
DSC	Differential scanning calorimetry
E, E', E'', E^*	Tensile modulus, its storage, loss and complex value
EPDM	Elastomeric terpolymer from ethylene, propylene and non-conjugated diene

EPR	Elastomeric copolymer of ethylene and propylene
EVAc	Ethylene–vinyl acetate copolymer
EVAl, EVOH	Ethylene–vinyl alcohol copolymer
G, G', G'', G^*	Shear modulus, its storage, loss and complex value
HDPE	High density polyethylene
HIPS	High impact polystyrene
LCP	Liquid crystal polymer
LD/LLDPE	Low density/linear low density polyethylene
m-PP	Acid or anhydride modified polypropylene; maleated polypropylene
MW, MWD	Molecular weight, molecular weight distribution
n	Exponent in Avrami's equation
NBR	Nitrile rubber
N_D	Deborah number
N_1	First normal stress difference
NDB	Negatively deviating blends
PABs	Polymer alloys and blends
PA	Polyamide
PB	Polybutadiene
PB-1	Polybutene-1
PC	Polycarbonate
PCL	Polycaprolactone
PCTG	Poly(cyclohexylenedimethylene-co-ethyleneterephthalate)
PDB	Positively deviating blends
PDS	Poly(styrene-co-2,2,6,6-tetramethyl-4-piperidinyl methacrylate)
PNDB	Positively and negatively deviating blends
PE	Polyethylene
PEI	Polyetherimide
PEST	Thermoplastic polyesters, e.g. PBT, PBT
PET	Polyethyleneterephthalate
PO	Polyolefin
PP	Isotactic polypropylene
PS	Polystyrene
PU	Polyurethane
RTPO	Reactor-made olefinic thermoplastic elastomer
SBS	Styrene–butadiene–styrene tri-block copolymer
SBR	Styrene–butadiene rubber
SEBS	Styrene–(ethylene/butylene)–styrene tri-block copolymer
SAN	Styrene–acrylonitrile copolymer
T, T_c, T_g, T_m	Temperature, crystallization, glass and melting temperature, respectively
T_d, T_{dr}	Degradation, drawing temperature
T_p	Processing temperature
TPO	Olefinic thermoplastic elastomer

t	Time
w, w_i	Weight fraction, weight fraction of component i
α	Geometrical model parameter in Takayanagi's Equation 9
β	Interlayer slip coefficient
$\dot{\gamma}$	Shear rate
δ	Geometrical model parameter in Takayanagi's Equation 9
Δ	Increment
η, η_0	Viscosity, zero shear rate viscosity
η_1, η_2	Viscosity of dispersed and matrix phase, respectively.
η_{max}	Viscosity enhancement due to the emulsion-like blend behavior
$\lambda=\eta_1/\eta_2$	Viscosity ratio.
$\kappa=d\sigma_{12}/v_{12}$	Capillarity (or Taylor) number.
σ_{12}	Shear stress.
$\sigma_{11}-\sigma_{22}$	The first normal stress difference.
v_{12}	Interfacial tension between phase 1 and 2.
ϕ, ϕ_m	Volume fraction, maximum packing volume fraction.
ϕ_p, ϕ_I	Volume fraction at the onset of the phase co-continuity and at the phase inversion, respectively
χ	Huggins–Flory binary thermodynamic interaction parameter

3.11 REFERENCES

1. Utracki, L. A. (1989) Melt flow of polyethylene blends, in *Multiphase Polymers: Blends and Ionomers*, (eds L. A. Utracki and R. A. Weiss) ACS Symposium Series, Vol. 395, Washington, DC, pp. 153–210.
2. Utracki, L. A. (1989) *Polymer Alloys and Blends*, Hanser-Verlag, Munich (also published by Tokyo Kagaku Dozin Co. Ltd., Tokyo, 1991).
3. Utracki, L. A. (ed.) (1991) *Two-Phase Polymer Systems*, Polymer Processing Society Progress in Polymer Processing Series, Vol. 2, Hanser, Munich.
4. Utracki, L. A. (1991) *Polymer Networks and Blends*, **1**, 61–9.
5. Utracki, L. A. (1991) *Progress in Polymer Processing, in Frontiers of Materials Research, Electronic and Materials Research*, (eds M. Kong and L. Huang) Elsevier, Amsterdam, pp. 39–48.
6. Olabisi, O., Robeson, L. M. and Shaw, M. T. (1979) *Polymer–Polymer Miscibility*, Academic Press, New York.
7. Lohse, D. J. (1986) *Polymer Engineering and Science*, **26**, 1500–9.
8. Lohse, D. J. and Wissler, G.E. (1991) *Journal of Materials Science*, **26**, 743–8.
9. Valenza, A., Lyngaae-Jørgensen, J., Utracki, L. A. and Sammut, P. (1991) *Polymer Networks and Blends*, **1**, 79–92.
10. Utracki, L. A. and Shi, Z.-H. (1992) *Polymer Engineering and Science*, **32**, 1824–33.

10a Shi, Z.-H. and Utracki, L. A. (1992) *Polymer Engineering and Science*, **32**, 1834–45.

11. Utracki, L. A. (1991) *Journal of Rheology*, **35**, 1615–37.
12. Tervoort-Engelen, Y. and Van Gisbergen, J. (1991) *Polymer Communications*, **32**, 261–3.

13. Rose, W. and Meurer, C. (1987) *Siemens Forsch.-u. Entwickl. – Ber. Bd.*, **16**, 127–30.
14. Bordereau, V., Shi, Z.-H., Utracki, L. A., Sammut, P. and Carrega, M. (1992) *Polymer Engineering and Science*, **32**, 1846–56.
15. Lin, C. C. (1979) *Polymer Journal*, **11**, 185–92.
16. Mao, L. J., Zhang, Z. P. and Ying, S. K. (1991) *Polymer Communications*, **32**, 242–4.
17. Nadkarni, V. M. and Jog, J. P. (1991) Crystallization behavior in polymer blends, in *Two-Phase Polymer Systems*, (ed. L. A. Utracki), Polymer Processing Society Progress in Polymer Processing Series, Vol. 2, Hanser, Munich, pp. 213–40.
18. Paul, D. R. and Barlow, J. W. (1979) *Div. Org. Coat. Plast. Chem. Prepr.*, **40**, 745–50.
19. Takayanagi, M., Uemura, S. and Minami, S. (1964) *Journal of Polymer Science, Part C*, **5**, 113–22.
20. Utracki, L. A. and Sammut P. (1991) *Plastics and Rubber Composites Processing and Applications*, **16**, 221–9.
21. Dumoulin, M. M. (1988) PhD Thesis, Ecole Polytechnique, Montreal.
22. Dumoulin, M. M. and Utracki, L. A. (1990) Time–temperature superposition for polyethylene/polypropylene blends, in *Rheology and Polymer Processing*, (eds A. A. Collyer and L. A. Utracki) pp. 207–40.
23. Plati, E. and Williams, J. G. (1975) *Polymer Engineering and Science*, **15**, 470–7.
24. Williams, J. G. (1977) *Polymer Engineering and Science*, **17**, 144–9.
25. Bucknall, C. B. (1977) *Toughened Polymers*, Applied Science, London.
26. Collyer, A. A. and Utracki, L. A. (eds) (1990) *Rheology and Polymer Processing*, Elsevier, Barking.
27. Han, C. D. (ed.) (1984) *Polymer Blends and Composites in Multiphase Systems*, American Chemical Society, Washington, DC.
28. Kleintjens, L. A. and Lemstra, P. J. (eds) (1985) *Integration of Fundamental Polymer Science and Technology*, Vol. 1, Elsevier, London; Lemstra, P. J. and Kleitjens, L. A. (eds) (1987–1991) *Integration of Fundamental Polymer Science and Technology*, Vols 2–5, Elsevier, London.
29. Klempner, D. and Frish, K. C. (eds) (1977, 1980, 1983) *Polymer Alloys*, Vol. I–III, Plenum Press, New York.
30. Manson, J. A. and Sperling, L. H. (1976) *Polymer Blends and Composites*, Plenum Press, New York.
31. Ottenbrite, R. M., Utracki, L. A. and Inoue, S. (eds) (1987) *Current Topics in Polymer Science*, Hanser-Verlag, Munich.
32. Paul, D. R. and Newman, S. (1978) *Polymer Blends*, Academic Press, New York.
33. Singh, A. and Silverman, J. (1992) (eds), *Radiation Processing of Polymers*, Polymer Processing Society Progress in Polymer Processing Series, Vol. 3, Hanser, Munich.
34. Šolc, K. (ed.) (1980, 1982) *Polymer Compatibility and Incompatibility*, Harwood Academic, New York (also published by MMI, New York, 1986).
35. Sperling, L. H. (1981) *Interpenetrating Polymer Networks and Related Materials*, Plenum Press, New York.
36. Klempner, D., Sperling, L. A. and Utracki, L. A. (eds) (1994) *Interpenetrating Polymer Networks*, American Chemical Society, Adv. Chem. Series No. 239, Washington, DC.
37. Utracki, L. A. (1987) Polymer blends: Their preparation, processing and economic forecast, in *Polymers and Composites*, Oxford University Press and I. B. Publishers, New Delhi.

38. Utracki, L. A. and Favis, B. (1988) Polymer alloys and blends, in *Encyclopedia of Engineering Materials*, Vol. 4, (ed. N. P. Cheremisinoff) Marcel Dekker, Matawan, NJ, pp. 121–201.
39. Utracki, L. A. and Weiss, R. A. (eds) (1989) *Multiphase Polymers: Blends and Ionomers*, ACS Symposium Series No. 395, Washington, DC.
40. Walsh, D. J., Higgins, J. S. and Maconnachie, A. (eds) (1985) *Polymer Blends and Mixtures*, NATO Series E, No. 89, Martinus Nijhoff, Dordrecht.
41. Galli, P., Foschini, G. and Moro, A. (1984) Ethylene–propylene rubber and polyolefin polymer blends: Present situation and future trends, in *International Symposium on Polymer Alloys: Structure and Properties*, Brugge, 4–7 June.
42. Galli, P., Danesi, S. and Simonazzi, T. (1984) *Polymer Engineering and Science*, **24**, 544–54.
43. Dumoulin, M. M., Carreau, P. J. and Utracki, L. A. (1987) *Polymer Engineering and Science*, **27**, 1627–33.
44. Dumoulin, M. M., Utracki, L. A. and Carreau, P. J. (1991) Melt rheology and morphology of linear low density polyethylene/polypropylene blends, in *Two-Phase Polymer Systems* (ed. L. A. Utracki), Polymer Processing Society Progress in Polymer Processing Series, Vol. 2, Hanser, Munich, pp. 188–212.
45. Xavier, S. F. (1991) Development of fine morphology in polypropylene composites, in *Two-Phase Polymer Systems* (ed. L. A. Utracki), Polymer Processing Society Progress in Polymer Processing Series, Vol. 2, Hanser, Munich, pp. 347–70.
46. Werner, R. J., Jennings, W. J., Crowley, D. L. and Smith, F. J. (1982) US Patent 4 366 292 (unassigned).
47. Dynamit Nobel AG. (1982) US Patent 4 319 004.
48. US Steel Corp. (1985) US Patent 4 514 534.
49. Sumitomo Chemical Co. Ltd. (1988) European Patent Application 285 414 A2.
50. Mobil Oil Corp. (1990) US Patent 4 929 681.
51. Amoco Corp. (1991) US Patent 5 041 491.
52. BASF AG (1991) European Patent Application 462 448 A2.
53. BASF AG (1991) European Patent Application 462 373.
54. Toyoda Gosei Co. Ltd (1991) German Patent 4 113 826 A1.
55. Cheung, P., Suwanda, D. and Balke, S. T. (1990) *Polymer Engineering and Science*, **30**, 1063–72.
56. Rizzo, G., Spadaro, G., Acierno, D. and Calderado, E. (1983) *Radiation and Physical Chemistry*, **21**, 349–53.
57. Van Gisbergen, J. and Overbergh, N. (1992) Radiation effects on polymer blends, in *Radiation Processing of Polymers* (eds A. Singh and J. Silverman), Polymer Processing Society Progress in Polymer Processing Series, Vol. 3 Hanser, Munich, pp. 51–70.
58. Lee, Y. K., Jeong, Y. T., Kim, K. C., Jeong, H. M. and Kim, B. K. (1991) *Polymer Engineering and Science*, **31**, 944–53.
59. Sawatari, C., Satoh, S. and Matsuo, M. (1990) *Polymer*, **31**, 1456–63.
60. Plochocki, A. P. (1978) Polyolefin blends: Rheology, melt mixing, and applications, in *Polymer Blends* (eds D. R. Paul and S. Newman), Academic Press, New York, pp. 319–68.
61. Plochocki, A. P. (1983) *Advances in Polymer Technology*, **2**, 267–81.
62. Valenza, A., La Mantia, F. P. and Acierno, D. (1984) *European Polymer Journal*, **20**, 727–31.
63. Nishimura, T. (1984) *Rheologica Acta*, **23**, 617–26.
64. Alle, N. and Lyngaae-Jørgensen, J. (1980) *Rheologica Acta*, **19**, 94–103.

65. Alle, N., Andersen, F. E. and Lyngaae-Jørgensen, J. (1981) *Rheologica Acta*, **20**, 222–30.
66. Fujiyama, M. and Kawasaki, Y. (1991) *Journal of Applied Polymer Science*, **42**, 467–80.
67. Fujiyama, M. and Kawasaki, Y. (1991) *Journal of Applied Polymer Science*, **42**, 481–8.
68. Chuu, M. S. and Tsai, B. C. (1991) Melt rheology and strain-induced crystallization of polypropylene and its high density polyethylene blends, in *Polymers as Rheology Modifiers*, (eds D. N. Schulz and J. E. Glass) ACS Symposium Series No. 462, Washington, DC, pp. 312–21.
69. Santamaria, A., Munoz, M. E., Pena, J. J. and Remiro, P. (1985) *Angewandte Makromolekulare Chemie*, **134**, 63–71.
70. Dumoulin, M. M., Farha, C. and Utracki, L. A. (1984) *Polymer Engineering and Science*, **24**, 1319–26.
71. Dumoulin, M. M., Utracki, L. A. and Carreau, P. J. (1988) *Rheologica Acta, Suppl.*, **26**, 215–7.
72. Dumoulin, M. M., Utracki, L. A. and Carreau, P. J. (1988) Melt rheology of LLDPE/PP blends: Dynamic behavior, in 4th Annual PPS Meeting, Orlando, FL, 8–11 May.
73. Palesek, M. and Malac, Z. (1986) Morphology of polymers, in Proceedings of the 17th Europhysics Conference on Macromolecular Physics, Prague, 15–18 July 1985, (ed. B. Sedlaček) W. de Gruyter & Co., Berlin.
74. Wenig, W. and Meyer, K. (1980) *Colloid and Polymer Science*, **258**, 1009–14.
75. Kuptsov, S. A., Erina, N. A. and Minina, O. D., Zhorin, V. A., Prut, E. V. and Antipov, E. M. (1991) *Vysokomolkulyarnie Soedineniya*, **B33**, 529–31.
76. Bartczak, Z., Galeski, A. and Martuscelli, E. (1984) Polymer Engineering and Science, **24**, 1155–65.
77. Bartczak, Z., Galeski, A. and Pracella, M. (1986) *Polymer*, **27**, 537–43.
78. Galeski, A., Pracella, M. and Martuscelli, E. (1984) *Journal of Polymer Science, Polymer Physics Edition*, **22**, 739–47.
79. Martuscelli, E., Pracella, M., Della Volpe, G. and Greco, P. (1984) *Makromolekulare Chemie*, **185**, 1041–61.
80. Bartczak, Z. and Galeski, A. (1986) *Polymer*, **27**, 544–8.
81. Noel, O. F. and Carley, J. F. (1975) *Polymer Engineering and Science*, **15**, 117–26;
82. Noel, O. F. and Carley, J. F. (1984) *Polymer Engineering and Science*, **24**, 488–92.
83. Lovinger, A. J. and Williams, M. L. (1980) *Journal of Applied Polymer Science*, **25**, 1703–13.
84. Nakafuku, C. (1983) *Polymer Journal*, **15**, 641–7.
85. Gallagher, G. A., Jakeways, R. and Ward, I. M. (1991) *Journal of Applied Polymer Science*, **43**, 1399–409.
86. Balta-Calleja, F. J., Santa Cruz, C., Sawatari, C. and Asano, T. (1990) *Macromolecules*, **23**, 5352–5.
87. Yeh, P.-L. and Birley, A. W. (1985) *Plastic and Rubber Processing and Applications*, **5**, 249–52.
88. Long, Y., Stachurski, Z. H. and Shanks, R. A. (1991) *Polymer International*, **26**, 143–6.
89. Gunderson, J. J. and Chilcote, S. M. (1991) *ACS Polymer Materials Science and Engineering*, **64**, 393–4.
90. Flaris, V., Wasiak, A. and Wenig, W. (1992) *SPE Technical Papers*, **38**, 1265–6.
91. Rizzo, G. and Spadaro, G. (1988) *European Polymer Journal*, **24**, 303–6.

92. Spadaro, G. and Rizzo, G. (1989) *European Polymer Journal*, **25**, 1189–92.
93. Bartlett, D. W., Barlow, J. W. and Paul, D. R. (1982) *Journal of Applied Polymer Science*, **27**, 2351–60.
94. Bartlett, D. W., Paul, D. R. and Barlow, J. W. (1981) *Modern Plastics*, December, 60–2.
95. Ramsteiner, F., Kanig, G., Heckmann, W. and Gruber, W. (1983) *Polymer*, **24**, 365–70.
96. Deanin, R. D. and D'Isidoro, G. E. (1980) *ACS Division of Polymer Chemistry Preprints*, **43**, 19–22.
97. Robertson, R. E. and Paul, D. R. (1973) *Journal of Applied Polymer Science*, **17**, 2579–95.
98. Deanin, R. D. and Sansone, M. F. (1978) *ACS Division of Polymer Chemistry Preprints*, **19**, 211–5.
99. Nolley, E., Barlow, J. W. and Paul, D. R. (1980) *Polymer Engineering and Science*, **20**, 364–9.
100. Du, Q. and Wang, L. (1989) *Journal of Polymer Science, Part B, Polymer Physics*, **27**, 581–8.
101. Greco, R., Mucciariello, G., Ragosta, G. and Martuscelli, E. (1981) *Journal of Materials Science*, **16**, 1001–7.
102. Rizzo, G. and Spadaro, G. (1984) *Polymer Engineering and Science*, **24**, 264–7.
103. Barlow, J. W. and Paul, D. R. (1984) *Polymer Engineering and Science*, **24**, 525–34.
104. Coran, A. Y., Patel, R. and Williams-Head, D. (1985) *Rubber Chemistry and Technology*, **58**, 1014–23.
105. Coran, A. Y. (1988) Thermoplastic rubber–plastic blends, in *Handbook of Elastomers* (eds A. K. Bhowmick and H. L. Stephens), Marcel Dekker, New York.
106. Shell Oil Co. (1991) European Patent Application, 461 883 A2.
107. Lohse, D. J., Datta, S. and Kresge, E. N. (1991) *Macromolecules*, **24**, 561–6.
108. Lohse, D. J. and Datta, S. (1992) *SPE Technical Papers*, **38**, 1036–8.
109. Hoechst AG (1992) German Patent, 4 021 899 A1.
110. Chodak, I., Janigova, I. and Romanov, A. (1991) *Makromolekulare Chemie*, **192**, 2791–9.
111. Leaversuch, R. D. (1992) *Modern Plastics*, **69**(4), 70–72.
112. Illinois Tool Works, Inc., (1991) European Patent Application, 462 722 A2.
113. Exxon Research and Engineering Co. (1981) US Patent, 4 251 646.
114. Ube Industries Ltd. (1982) US Patent, 4 363 885.
115. Phillips Petroleum Co. (1984) US Patent 4 489 187.
116. Exxon Research and Engineering Co. (1983) European Patent Application, 0 092 318.
117. Himont Inc. (1990) European Patent Application 0 373 660.
118. Chu, C. Y. (1992) *SPE Technical Papers*, **38**, 1664–9.
119. Bhardwaj, I. S. (1990) Development and characterization of ultrahigh impact polyolefin blends, in 33rd IUPAC International Symposium on Macromolecules, paper 1183, Montreal, Canada, 8–13 July.
120. Agarwal, P. K., Makowski, H. S. and Lundberg, R. D. (1980) *Macromolecules*, **13**, 1679–86.
121. Duvdevani, I., Agarwal, P. K. and Lundberg, A. D. (1982) *Polymer Engineering and Science*, **22**, 499–506.
122. Kresge, E. N. (1991) *Rubber Chemistry and Technology*, **64**, 469–79.
123. Romankevich, O. V. and Frenkel, S. Ja. (1980) *Acta Polymerica*, **31**, 287–92.
124. Plochocki, A. P. (1984) *Advances in Polymer Technology*, **3**, 405–10.

125. Lee, Y.-D. and Lu, C.-C. (1982) *Journal of the Chinese Institute of Chemical Engineering*, **13**, 1–8.
126. Chaudhary, V., Varma, H. S. and Varma, I. K. (1991) *Polymer*, **32**, 2534–45.
127. Van Gisbergen, J. G. M., M. Hoeben, W. F. L. and Meijer, H. E. H. (1991) *Polymer Engineering and Science*, **31**, 1539–44.
128. Karger-Kocsis, J. and Kiss, L. (1987) *Polymer Engineering and Science*, **27**, 254–62.
129. Pukánszky, B., Tüdős, F., Kalló, A. and Bodor, G. (1990) *Polymer*, **30**, 1399–407.
130. Pukánszky, B., Tüdós, F., Kalló, A. and Bodor, G. (1990) *Polymer*, **30**, 1407–13.
131. Karger-Kocsis, J., Kalló, A. and Kuleznev, V. N. (1981) *Acta Polymerica*, **32**, 578–81.
132. Karger-Kocsis, J. and Csikai, I. (1987) *Polymer Engineering and Science*, **27**, 241–53.
133. Fortelny, I., Kovar, J., Sikora, A., Hlavata, D., Krulis, Z., Novakova, Z., Pélzbaver, Z., and Cefelin, P. (1985) *Angewandte Makromolekulare Chemie*, **132**, 111–22.
134. Takahashi, T., Mizuno, H. and Thomas, E. L. (1983) *Journal of Macromolecular Science, Physics*, **B22**, 425–36.
135. Hoppner, D. and Wendorff, J. H. (1990) *Colloid and Polymer Science*, **268**, 500–12.
136. Pukánszky, B., Fortelny, I., Kovar, J. and Tüdős, F. (1991) *Plastic and Rubber Composite Processing and Applications*, **15**, 31–8.
137. Bartczak, Z., Galeski, A., Martuscelli, E. and Janik, H. (1985) *Polymer*, **26**, 1843–8.
138. D'Orazio, L., Mancarella, C., Martuscelli, E. and Sticotti, G. (1991) *Journal of Materials Science*, **26**, 4033–47.
139. D'Orazio, L., Mancarella, C., Martuscelli, E. and Polato, F. (1991) *Polymer*, **32**, 1186–94.
140. Yang, D., Zhang, B., Yang, Y., Fang, Z., Sun, G. and Feng, Z. (1984) *Polymer Engineering and Science*, **24**, 612–7.
141. Gupta, A. K. and Ratnam, B. K. (1991) *Journal of Applied Polymer Science*, **42**, 297–315.
142. Wenig, W. and Schoeller, T. (1991) *Angewandte Makromolekulare Chemie*, **192**, 183–97.
143. Siegmann, A. (1982) *Journal of Applied Polymer Science*, **27**, 1053–65.
144. Lee, M.-S. and Chen, S.-A. (1987) *Journal of Polymer Science, Part C: Polymer Letters*, **25**, 37–43.
145. Martuscelli, E., Silvestre, C. and Abate, G. (1982) *Polymer*, **23**, 229–37.
146. Wenig, W. and Fiedel, H.-W. (1991) *Makromolekulare Chemie*, **192**, 85–99.
147. Martuscelli, E., Canetti, M., Bonfatti, A. M. and Seves, A. (1991) *Polymer*, **32**, 641–7.
148. Marcandalli, B., Testa, G., Seves, A. and Martuscelli, E. (1991) *Polymer*, **32**, 3376–80.
149. Cimmino, S., Guarrata, P., Martuscelli, E., Silvestre, C. and Pier, P. (1991) *Polymer*, **32**, 3299–304.
150. Kausch, H. H. (1978) *Polymer Fracture*, Springer-Verlag, Berlin.
151. Karger-Kocsis, J. and Kuleznev, V. N. (1982) *Polymer*, **23**, 699–705.
152. Chompff, A. J. and Naderi, A. (1978) *ACS Division of Organic Coatings and Plastic Chemistry*, **38**, 700–4.
153. Kikuchi, Y., Fukui, T., Okada, T. and Inoue, T. (1991) *Polymer Engineering and Science*, **31**, 1029–32.

154. Kolarik, J., Agrawal, G. L., Krulis, Z. and Kovar, J. (1986) *Polymer Composites*, **7**, 463–71.
155. Kolarik, J., Vele, G., Agrawal, G. L. and Fortelny, I. (1986) *Polymer Composites*, **7**, 472–9.
156. Zhu, G., Wang, M., Cai, H., Yin, L. and Chen, S. (1991) Studies on morphology and properties of polypropylene materials used for automobile parts, in *Frontiers of Materials Research, Electronic and Materials Research*, Vol. 3 (eds H. Feng, Y. Han and L. Huang) Elsevier, Amsterdam, pp. 149–54.
157. Coppola, F., Greco, R., Martuscelli, E., Kammer, H. W. and Kummerlöwe, C. (1987) *Polymer*, **28**, 47–56.
158. Ramsteiner, F. (1991) *Acta Polymerica*, **42**, 584–9.
159. Ito, J., Mitani, K. and Mizutani, Y. (1984) *Journal of Applied Polymer Science*, **29**, 75–87.
160. Varga, J. and Garzó, G. (1990) *Angewandte Makromolekulare Chemie*, **180**, 15–33.
161. Fernando, P. L. and Williams, J. G. (1980) *Polymer Engineering and Science*, **20**, 215–20.
162. Fernando, P. L. and Williams, J. G. (1981) *Polymer Engineering and Science*, **21**, 1003–9.
163. Hodgkinson, J. M., Savadori, A. and Williams, J. G. (1983) *Journal of Materials Science*, **18**, 2319–36.
164. Jang, B. Z., Uhlmann, D. R. and Van der Sande, J. B. (1985) *Journal of Applied Polymer Science*, **30**, 2485–504.
165. Gupta, A. K., Srinivasa, K. R. and Kumar, P. K. (1991) *Journal of Applied Polymer Science*, **43**, 451–62.
166. Theisen, A., Hess, M. and Kosfeld, R. (1991) Analysis of ternary composites of polypropylene/EPDM/filler, in *Integration of Fundamental Polymers Science and Technology*, Vol. 5 (eds P. J. Lemstra and L. A. Kleitjens), Elsevier, London, pp. 72–6.
167. Ide, F. and Hasegawa, A. (1974) *Journal of Applied Polymer Science*, **18**, 963–74.
168. Standard Oil Co. (1983) US Patent 4 383 084.
169. Hobbs, S. Y., Bopp, R. C. and Watkins, V. H. Watkins, (1983) *Polymer Engineering and Science*, **23**, 380–9.
170. Deanin, R. D., Orroth, S. A. and Bhagat, R. I. (1990) *Polymer-Plastic Technology and Engineering*, **29**, 289–95.
171. Atochem SA (1988) French Patent Application 003 877.
172. ICI (1975) UK Patent 1 403 797.
173. DuPont (1983) US Patent, 4 410 482.
174. DuPont (1984) US Patent, 4 444 817.
175. Valenza, A., La Mantia, F. P., Demma, G. B., Romano, V. and Aciero, D. (1991) *Polymer Networks and Blends*, **1**, 71–7.
176. Liu, Q., Bronk, J. M., Verma, A. and Riffle J. S. (1991) *Polymer Materials Science and Engineering*, **64**, 158–60.
177. Hosoda, S., Kojima, K., Kanada, Y. and Aoyagi, M. (1991) *Polymer Networks and Blends*, **1**, 51–9.
178. Nishio, T., Suzuki, Y., Kojima, K. and Kakugo, M. (1990) *Kobunshi Robunshu*, **47**, 331–7.
179. Costin, R. and Billet, C. (1990) Novel thermoplastic impact modifiers for engineering thermoplastics, in Proceedings of Compalloy 90, New Orleans, 7–9 March, pp. 293–309.
180. Utracki, L. A. and Sammut, P. (1992) *Polymer Networks and Blends*, **2**, 23–39.
181. Dumoulin, M. M., Toma, P., Utracki, L. A., Jinnah, I. and Kamal, M. R. (1985) *SPE Technical Papers*, **31**, 534–6.

182. Dumoulin, M. M., Toma, P., Utracki, L. A., Jinnah, I. and Kamal, M. R. (1986) *Polymer Engineering and Science* **26** 34–44.
183. Liang, B.-R., White, J. L., Spruiell, J. E. and Goswami, B. C. (1983) *Journal of Applied Polymer Science*, **28**, 2011–32.
184. Yakovlev, K. V., Zhitomirets, R. I., Romankevich, O. V., Zabello, S. E. and Yudin, A.V. (1984) *Khimicheskaia Tekhnologia*, **5**, 14–16.
185. Yakovlev, K. V., Shevchenko, A. S., and Romankevich, O. V. (1986) *Ukrainskii Khimicheskii Zhurnal*, **52**, 771–5.
186. Yakovlev, K. V., Shevchenko, A. S. and Romankevich, O. V. (1986) *Soviet Progress in Chemistry*, **57**, 107–11.
187. Park, S. J., Kim, B. K. and Jeong, H. M. (1990) *European Polymer Journal*, **26**, 131–6.
188. Hayashida, K. and Yoshida, T. (1979) *Bulletin of the Faculty of Textile Science, Kyoto University*, **9**, 65–72.
189. Willis, J. M. and Favis, B. D. (1988) *Polymer Engineering and Science*, **28**, 1416–26.
190. Chen, C.-C., Fontan, K., Min, K. and White, J. L. (1988) *Polymer Engineering and Science*, **28**, 69–80.
191. Utracki, L. A. and Sammut, P. (1990) Rheological response of polyamide/polypropylene blends, in Polymer Processing Society Annual Meeting, Nice, France, 17–20 April.
192. Van Gheluwe, P., Favis, B. D. and Chalifoux, J.-P. (1988) *Journal of Materials Science*, **23**, 3910–20.
193. Lyngaae-Jørgensen, J., Thomsen, L. D., Rasmunssen, K., Søndergaard, K. and Andersen, F. E. (1988) *International Polymer Processing*, **2**, 123–30.
194. Utracki, L. A., Sammut, P., Lyngaae-Jørgensen, J. and Valenza, A. (1991) *SPE Technical Papers*, **37**, 978–81.
195. Scholz, P., Froelich, D. and Muller, R. (1989) *Journal of Rheology*, **33**, 481–499.
196. Chuang, H.-K. and Han, C. D. (1984) American Chemical Society, *Adv. Chem. Series No.* **206**, 171–83.
197. Chuang, H.-K. and Han, C. D. (1984) *SPE Technical Papers*, **30**, 451–4.
198. Rogers, M. and Samurkas, T. (1992) *Polymer Engineering and Science*, **32**, 1727–33.
199. Liang, B.-R., White, J. L., Spruiell, J. E. and Goswami, B. C. (1983) *Journal of Applied Polymer Science*, **28**, 2011–32.
200. Romankevich, O. V., Yakovlev, K. V., Shtanko, L. V. and Muzyka, S. A. (1986) *Khimicheskie Volokna*, **3**, 14–17.
201. Wajs, M. (1990) Improved properties, applications and markets for compatibilized polypropylene/nylon alloys, in Proceedings Compalloy Europe '90, Brussels, 8–9 October, pp. 145–54.
202. Diaz, M. I. A. and Maeda, Y. (1990) *Reports on Progress in Polymer Physics, Japan*, **33**, 255–58.
203. Maeda, Y. (1991) *ACS Polymer Chemistry Division, Polymer Preprints*, **32**(3) 281–2.
204. Todoki, M., Hatakeyama, T., Maeda, Y., Kanetsuna, H., Shimada, T. and Akiyama, S. (1991) Dynamic viscoelastic properties of immiscible polymer blends, in International Symposium on Prestandard Research for Advanced Materials, Tokyo, 16–18, December.
205. Sato, S., Li, Y., Sakairi, Y., Urano, Y. and Oyanagi, Y. (1989) *Kogakuin Daigaku Kenkyu Hokoku*, **67**, 36–40.
206. Grof, I., Sain, M. M. and Durcova, O. (1992) *Journal of Applied Polymer Science*, **44**, 1061–8.

207. Wippler, C. (1992) *Polymer Engineering and Science*, **32**, 347–52.
208. Liang, Z. and Williams, H. L. (1992) *Journal of Applied Polymer Science*, **44**, 699–717.
209. Adams, G. C. and Wu, T. K. (1986) Materials characterization by instrumented impact testing, in *Failure of Plastics* (eds W. Brostow and R. D. Corneliussen) Hanser, Munich, pp. 144–68.
210. Dobkowski, Z. (1980) *Polimery*, **25**, 110–13.
211. Idemitsu Petrochemical Co. (1991) European Patent Application 0 417 912 A2.
212. GE Plastics, (1984) European Patent Application 0 111 851.
213. GE Plastics (1985) US Patent, 4 520 164.
214. Mobil Oil Co. (1986) US Patent 4 568 723.
215. General Electric Co. (1989) US Patent, 4 806 597.
216. Tonen Sek. KK (1989) European Patent Application 0 308 179.
217. Dow Chemical Co. (1989) US Patent, 4 806 599.
218. Bayer AG (1991) German Patent, 4 009 759 A1.
219. Liang, Z. and Williams, H. L. (1991) *Journal of Applied Polymer Science*, **43**, 379–92.
220. Acierno, D., Demma, G., La Mantia, F. P., Martuscelli, E., Romano, V. and Valenza, A. (1990) *European Polymer Journal*, **26**, 1049–54.
221. Utracki, L. A. and Sammut, P. (1990) *Polymer Engineering and Science*, **30**, 1027–40.
222. Sakai, H., Kitano, T. and Nishimura, T. (1991) *Kobunshi Robunshu*, **48**, 81–8.
223. Sakai, H., Nishimura, T. Yoshimura, K., Todoki, M., Maehara, A., Matsunga, T. and Nakayama, K. (1990) *Kobunshi Robunshu*, **47**, 307–13.
224. Wippler, C. (1990) *Polymer Engineering and Science*, **30**, 1106–13.
225. Tomlins, P. E. and Read, B. E. (1991) *Plastic and Rubber Composites Processing and Applications*, **16**, 17–23.
226. Ajji, A. (1989) *Polymer Engineering and Science*, **29**, 1544–50.
227. Favis, B. D. and Chalifoux, J. P. (1988) *Polymer*, **29**, 1761–7.
228. Favis, B. D. and Therrien, D. (1991) *Polymer*, **32**, 1474–81.
229. Fisa, B., Favis, B. D. and Bourgeois, S. (1990) *Polymer Engineering and Science*, **30**, 1051–5.
230. BASF AG (1976) US Patent, 3 937 757.
231. Bayer AG (1983) US Patent, 4 395 517.
232. Anonymous (1984) *Research Disclosure*, **239**, 97.
233. Shell (1982) UK Patent Application 2 122 627A.
234. Rudin, A., Loucks, D. A. and Goldwasser, J. M. (1980) *Polymer Engineering and Science*, **20**, 741–6.
235. Liu, Y. (1991) *Fangzhi Xuebao*, **12**, 252–5.
236. Ye, Y.-C., La Mantia, F. P., Valenza, A., Citta, V., Pedretti, U. and Roggero, A. (1991) *European Polymer Journal*, **27**, 723–7.
237. Blizard, K. G., Haghighat, R. and Lusignea, R. W. (1990) Applications for extruded thermotropic liquid crystalline polyester blends, in Proceedings Compalloy Europe '90, Brussels, 8–9, October, pp. 295–309.
238. Anonymous (1990) *Plastics Brief*, **14**,(29).
239. Sukhadia, A. M., Datta, A. and Baird, D. G. (1991) *SPE Technical Papers*, **37**, 1008–13.
240. O'Donnell, H. J., Datta, A. and Baird, D. G. (1992) *SPE Technical Papers*, **38**, 2248–52.
241. Han, C. D., Villamizar, C. A., Kim, Y. W. and Chen, S. J. (1977) *Journal of Applied Polymer Science*, **21**, 353–70.

242. Rudin, A. and Brathwaite, N. E. (1984) *Polymer Engineering and Science*, **24**, 1312–18.
243. Mobil Oil Corp. (1987) US Patent 4 673 711.
244. Montefine SA (1982) European Patent Application 0 060 524.
245. Toyota Jidosha KK (1990) European Patent Application 0 376 081.
246. Phillips Petroleum Co. (1981) US Patent 4 258 145.
247. Mobil Oil Corp. (1987) European Patent Application 0 238 756.
248. RJR Archer, Inc. (1982) European Patent Application 0 045 580.
249. Miroshnikov, Yu. P. and Williams, H. L. (1982) *Vysokomolekulyarnie Soedineniya*, **A24**, 1594–605.
250. Miroshnikov, Yu. P. (1984) *Mekhanika Kompozitnykh Materialov*, 104–9.
251. Miroshnikov, Yu. P. and Andreyeva, E. N. (1987) *Vysokomolekulyarnie Soedineniya*, **A29**, 579–82.
252. Cohen, A. and Schroeder, J. R. (1990) *Journal of Rheology*, **34**, 685–704.
253. Deri, F., Genillon, R. and May, J. F. (1985) *Angewandte Makromolekulare Chemie*, **134**, 11–22.
254. Gupta, A. K. and Purwar, S. N. (1984) *Journal of Applied Polymer Science*, **29**, 1079–93.
255. Gupta, A. K. and Purwar, S. N. (1984) *Journal of Applied Polymer Science*, **29**, 1595–609.
256. Gupta, A. K. and Purwar, S. N. (1984) *Journal of Applied Polymer Science*, **29**, 3513–31.
257. Gupta, A. K. and Purwar, S. N. (1985) *Journal of Applied Polymer Science*, **30**, 1777–798.
258. Gupta, A. K. and Purwar, S. N. (1984) *Journal of Applied Polymer Science*, **29**, 1079–93, 1595–609, 3513–31.
259. Gupta, A. K. and Purwar, S. N. (1985) *Journal of Applied Polymer Science*, **30**, 1799–814.
260. Gupta, A. K. and Purwar, S. N. (1986) *Journal of Applied Polymer Science*, **31**, 535–51.
261. Maiti, S. N., Agarwal, V. and Gupta, A. K. (1991) *Journal of Applied Polymer Science*, **43**, 1891–900.
262. Van Oene, H. (1972) *Journal of Colloid and Interface Science*, **40**, 448–67.
263. Taylor, G. I. (1932) *Proceedings of the Royal Society*, **A138**, 41–8.
264. Taylor, G. I. (1934) *Proceedings of the Royal Society*, **A146**, 501–23.
265. Fan, Y. and Hou, X. (1982) *Kao Fen Tzu Tung Hsu*, **5**, 343–8.
266. Verbraak, C. L. J. A., Van Dam, J. and Van der Vegt, A. K. (1985) *Polymer Engineering and Science*, **25**, 431–3.
267. Wenig, W., Fiedel, H.-W. and Scholl, A. (1990) *Colloid and Polymer Science*, **268**, 528–35.
268. Xu, X., Xia, Z. and Li, H. (1991) A study on morphologies and structures of polypropylene (PP)/SBR blends, in *Frontiers of Materials Research, Electronic and Materials Research*, Vol. 3 (eds H. Feng, Y. Han and L. Huang) Elsevier, Amsterdam, pp. 79–84.
269. Hlavata, D., Plestil, J., Żuchowska, D. and Steller, R. (1991) *Polymer*, **32**, 3313–16.
270. Shi, Z.-H., Sammut, P., Bordereau, V. and Utracki, L. A. (1992) *SPE Technical Papers*, **38**, 1818–21.
271. Markin, C. and Williams, H. L. (1980) *Journal of Applied Polymer Science*, **25**, 2451–66.
272. Petrovic, Z., Budinski-Simendic, J., Nemes, A. and Divjakovic, V. (1988) Polypropylene–Polyurethane Blends, in Proceedings of EUROTECH, European Technical Conference on PP, Paris, pp. 7.0–7.9.

273. Bánhegyi, G., Karasz, F. E. and Petrovic, Z. S. (1990) *Journal of Applied Polymer Science*, **40**, 435–52.
274. Petrovic, Z., Budinski-Simendic, J. and Divjakovic, V. (1991) *Journal of Applied Polymer Science*, **42**, 779–90.
275. Monsanto Co. (1981) European Patent Application 0 036 279 A3.
276. Hercules Inc. (1982) Republic of South Africa Patent Application 103 025.
277. Du Pont de Nemours and Co. (1985) European Patent Application 0 136 059.
278. General Electric Co. (1984) US Patent 4 427 830.
279. Kalfoglou, N. K. (1985) *Journal of Applied Polymer Science*, **30**, 1989–97.
280. Baitoul, M., Saint-Guiron, H., Xans, P. and Monge, Ph. (1983) *European Polymer Journal*, **19**, 651–65.

4

Structure and properties of polypropylene–elastomer blends*

E. Martuscelli

4.1 INTRODUCTION

For a given polypropylene (PP)/elastomer composition, the melt rheological behaviour, the phase structure in the melt and, after crystallization, the mode and state of dispersion of components, the crystalline structure and texture, and the adhesion between matrix and dispersed domains are the key parameters. The properties of PP/elastomer blends therefore depend on the following factors related to [1, 2]:

1. the molecular characteristics of both components (constitution, configuration, molecular mass and molecular mass distribution);
2. the blending, processing and crystallization conditions.

A comprehensive review concerning the influence of these factors on the structure and properties of PP/elastomer blends is reported in this chapter.

4.2 INFLUENCE OF THE MOLECULAR STRUCTURE, OF THE RUBBERY COMPONENTS

4.2.1 Influence of molecular mass

Polypropylene–polyisobutylene (PP/PIB) system

A pioneering experimental study was carried out to investigate the kinetic, morphological and thermodynamic properties of thin films, obtained by

*Further information on these topics can be found in Chapter 7 Volume 1, Chapter 3 Volume 3 and Chapter 4 Volume 3.

Polypropylene: Structure, blends and composites. Edited by J. Karger-Kocsis.
Published in 1995 by Chapman & Hall, London. ISBN 0 412 58430 1

blending PP with three samples of polyisobutylene (PIB) with different molecular mass [3]. PIB is an elastomer whose repeat unit has the molecular structure:

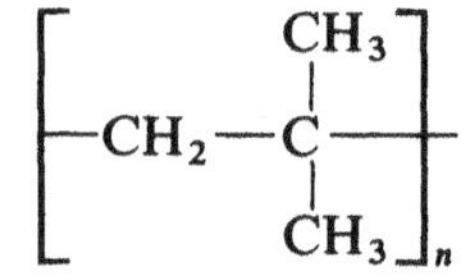

The following procedure was used for blending and film preparation:

1. The components (see Table 4.1 for molecular characteristics) were dissolved in xylene at 120 °C.
2. The solvent was rapidly evaporated and the resultant powder dried.
3. The blend powders were compression moulded to thin films at about 200 °C.

Table 4.1 Molecular characteristics, sources and trade name of PP and PIB samples used [3]

Polymer	*Source and trade name*	*Molecular mass (g/mol)*	*Code*
Isotactic poly-propylene	RAPRA	$\bar{M}_w = 3.07 \times 10^5$ $\bar{M}_n = 1.56 \times 10^4$	PP
Polyisobutylene	Vistanex LM MH (Esso)	$\bar{M}_v = 6.6 \times 10^4$	PIB (LM)
	Vistanex L120	$\bar{M}_v = 1.6 \times 10^6$	PIB (MM)
	EGA Chemie	$\bar{M}_v = 3.5 \times 10^6$	PIB (HM)

Optical microscopy was used to study the development of phase structure during crystallization, the mode and state of dispersion of the minor component (PIB) and the radial growth rate of PP spherulites as a function of the molecular mass of PIB and the crystallization temperature [3–5].

Analysis of optical micrographs of thin films of PP/PIB, isothermally crystallized at relatively low undercooling (ΔT), shows that the mode and state of dispersion of PIB is strongly dependent, for a given composition, on the molecular mass of PIB [3–5]. The results of this investigation are summarized in the following subsections.

PP/PIB(LM) system

At low PIB concentration almost all elastomer, for all values of T_c studied, seems to be rejected between PP spherulites. As the content of PIB increases some tiny drops of low molecular mass PIB (PIB(LM)) are trapped inside the spherulites. The higher the concentration, the more particles are occluded (Figure 4.1) [3]. The spherulite growth rate (G) shows, as a function of PIB(LM) content, a rather significant decrease at low concentration followed by an almost constant trend. For relatively higher values of ΔT a minimum is observed (Figure 4.2) [4, 5]. This behaviour may suggest that in the melt the two components are miscible, but phase separated at the crystallization temperature (T_c).

PP/PIB(MM) system

In all films with medium molecular mass PIB (PIB(MM)) distinct separated amorphous drops of PIB are observed. During crystallization these PIB particles are rejected in all cases into the spherulite boundaries [4, 5]. The particles, once occluded, undergo coalescence. The development of phase structure in the melt and in the solid following PP crystallization is clearly shown by Figure 4.3, where the crystallized spherulite of PP in the blend melt and the phase structure of the melt after melting of the spherulite are seen [4, 5].

The decreasing trend observed, at constant T_c, in plots of G versus PIB content (Figure 4.4) was explained assuming that during crystallization a certain amount of energy must be spent by the crystallizing field to reject in the melt the pre-existing dispersed particles of PIB [4, 5]. According to the theory elaborated by Martuscelli and colleagues [4, 5] this energy represents a new energy barrier affecting the growth rate of spherulites in PP/PIB blends.

Thus the equation for G, in the case of blends where spherulites grow in the presence of dispersed domains that are initially rejected by the growing front, is:

$$G = G_1 \exp(-E_1)/KT \tag{1}$$

where $G_1 = G_0 \exp(-\Delta F^*/KT) \exp(-\Delta\Phi^*/KT)$ is the expression of G for a plain polymer and E_1 is the energy dissipated by the growing front for rejection of drops pre-existing in the melt. Expressions for the possible dissipation energy terms and for the corresponding growth rate of spherulites in a crystallizable–non-crystallizable polymer–polymer system were derived by Bartczak, Galeski and Martuscelli [4]. The energy terms were estimated in the case of isotactic (iPP)/PP rubber blends for which the radial spherulite growth rate as a function of the composition and crystallization temperature was already determined [3–5]. The results are shown in Table 4.2. From the values of Table 4.2 it was concluded that only rejection may have a significant influence on the spherulite growth rate of PP [4, 5].

PP/PIB(HM) system

In films of PP/PIB(HM) (90/10) blend the elastomer is mainly ejected at the spherulite boundaries and probably also in the interlamellar regions,

Figure 4.1 Optical micrographs of melt crystallized films of iPP/PIB (LM) blends, $T_c = 131$ °C, crossed polarizers: (a) 90/10, (b) 80/20, (c) 60/40 [3].

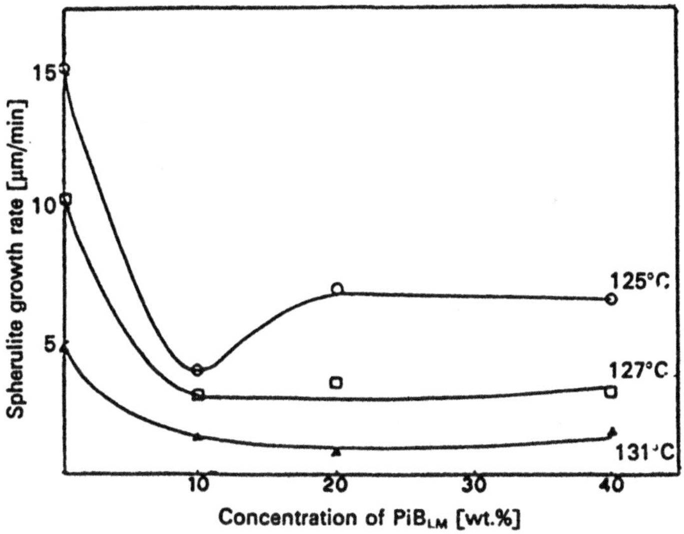

Figure 4.2 Spherulite growth rate of PP as a function of PIB(LM) content at different crystallization temperatures [4, 5].

since the spherulites at the border are more open and coarse [3] (Figure 4.5a).

It was observed that when the content of PIB(HM) increased a drop-like structure appeared in the melt. During crystallization these drops are occluded after rejection and some of them coalesce (Figure 4.5b).

As T_c increases more material is rejected into the interspherulitic regions and less remains inside the spherulite [4, 5]. At higher PIB(HM) content (60/40) an amorphous continuous phase permeated by crystalline PP spherulites is observed (Figure 4.5c) [3].

The growth rate of PP spherulites shows a large decrease for low PIB content followed by a slight increase (Figure 4.6) [4, 5]. Based on the above comparison of the microstructure of blends containing PIB with different molecular mass, it emerges that after complete crystallization, at a given T_c, the phase structure is drastically dependent on the molecular mass of PIB (compare Figures 4.1, 4.5 and 4.7).

It is obvious, for example, that for a 80/20 blend composition, larger dispersed drop-like particles are observed in the blend film containing PIB with higher molecular mass (compare Figures 4.1b, 4.5b and 4.7b) [3].

The melting behaviour of PP/PIB blend films also turns out to be dependent upon the molecular mass of elastomer as shown by the trend of the curves of Figure 4.8, where the apparent melting temperatures, T'_m are plotted, at a given T_c, as function of PIB content. The depression observed in T'_m is probably related to both diluent and morphological effects [3]. Some of the results may be accounted for by assuming that under certain

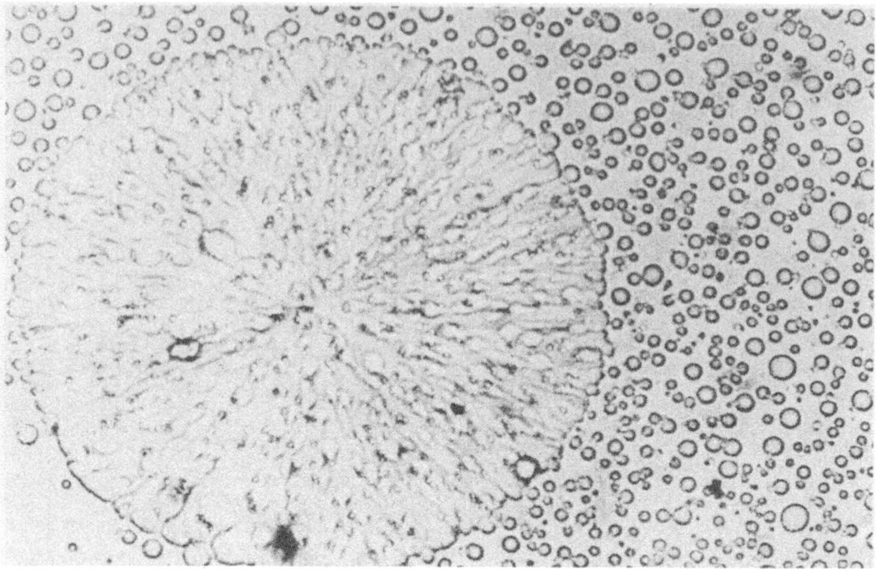

(a)

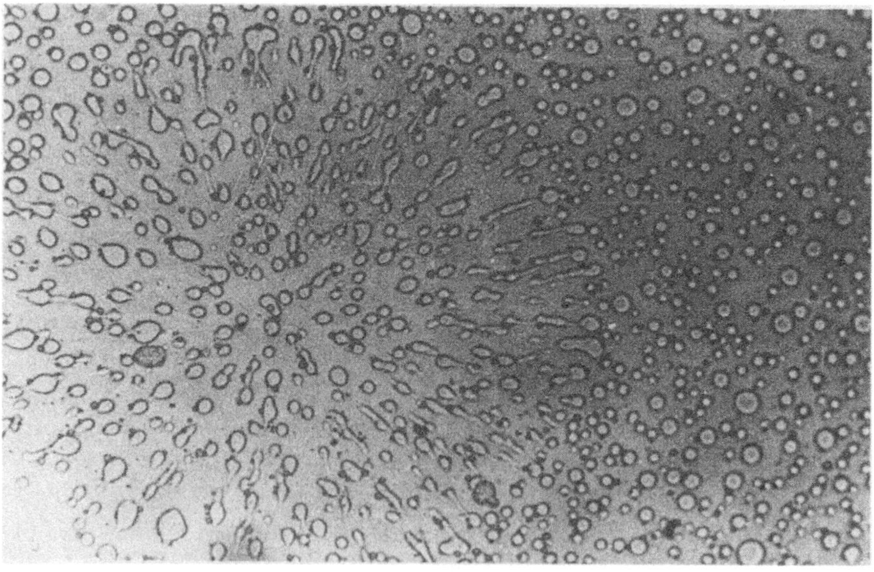

(b)

Figure 4.3 Optical micrographs of PP/PIB(MM) (80/20) blend film; crystallization temperature 133 °C: (a) before and (b) after PP spherulite melting [4, 5].

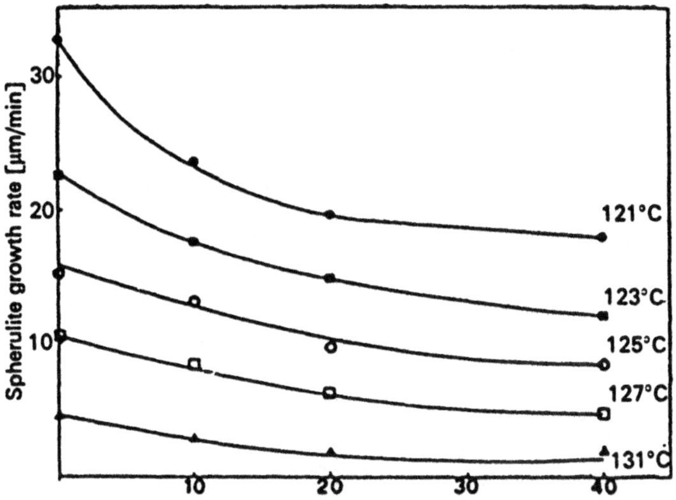

Figure 4.4 Spherulite growth rate of PP as a function of PIB(MM) content at various crystallization temperatures [4, 5].

Table 4.2 Energies dissipated by the crystallizing PP spherulites to perform rejection, occlusion and deformation of dispersed particles pre-existing in the melt as separated phase [4, 5]

Process	*Energy* (J/mol *of PP repeating units*)
Rejection	10^{+1} –10^{+4}
Kinetic energy of rejection	10^{-15}–10^{-4}
Occlusion	10^{-2} –10^{-1}
Deformation (only the surface change considered)	10^{-1} –10^{0}
Viscous term	10^{-7} –10^{-6}

conditions and for some molecular mass, PP and PIB may be miscible or phase separated in the melt. The hypothesis that PP and PIB may be miscible under certain conditions is supported by the comparison of the solubility parameters of the two polymers [3]. The solubility parameter, δ, is usually defined as the square root of the vaporization energy per unit volume. The probability for two polymers to be miscible is high if they have close values of δ.

The solubility parameter was calculated by using the following equation:

$$\delta = \rho \frac{\sum F_i}{M} \tag{2}$$

where ρ is the density of the polymer at the reference temperature, M is the molecular mass of the repetitive units of the polymer and ΣF_i is the sum of

(a)

(b)

(c)

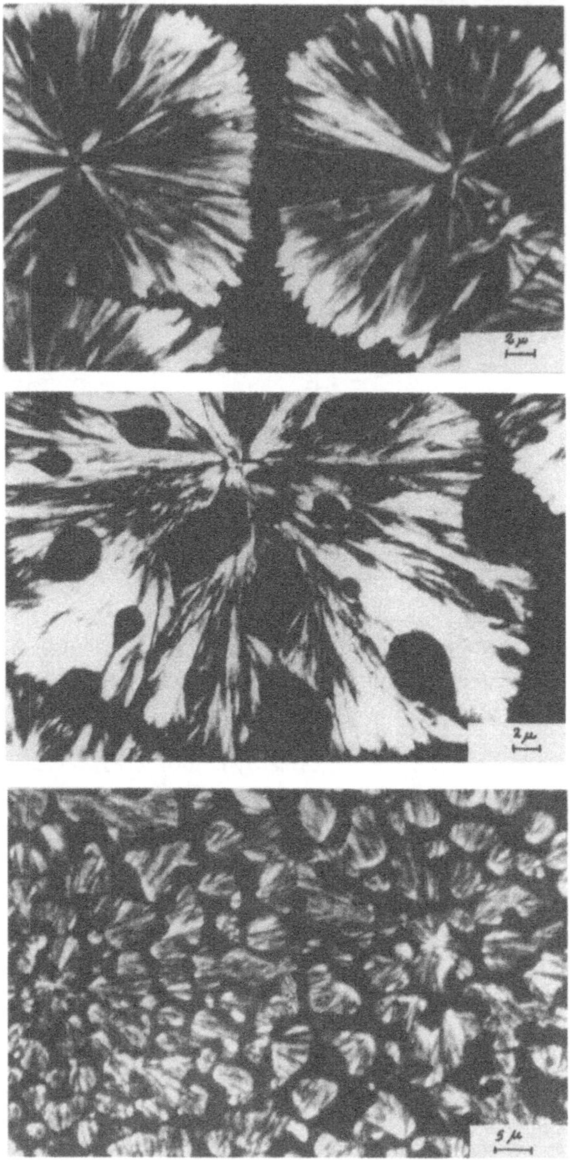

Figure 4.5 Optical micrographs of melt crystallized films of PP/PIB(HM) blends. $T_c = 131\,°C$, crossed polarizers: (a) 90/10, (b) 80/20, (c) 60/40 [3].

the molar attraction constants of all the chemical groups of the repetitive units [3]. Using literature equations for the temperature dependence of the density of PIB and PP amorphous phase, and calculating the attraction constant by means of the Hoy table [3], it was possible to calculate the values of δ as a function of T for the two polymers.

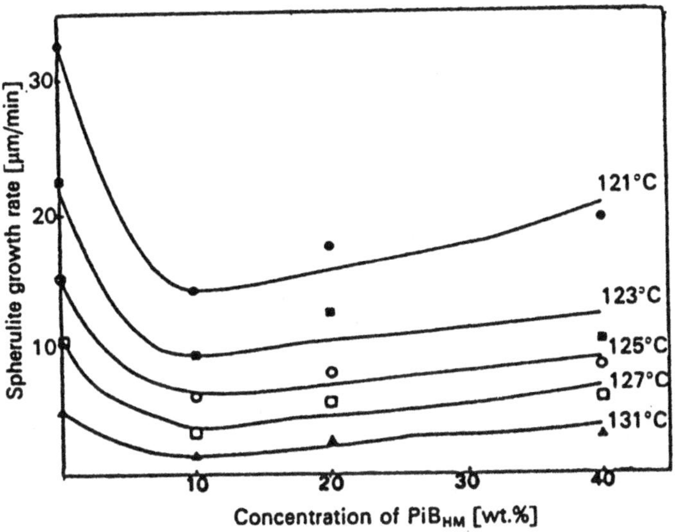

Figure 4.6 Spherulite growth rate of PP as a function of PIB(HM) content at various temperatures [4, 5].

The δ values for PIB and PP were found to be very close [3], with a difference of about 0.41 $(J/cm^3)^{1/2}$ in the 100–200 °C range (Figure 4.9).

This result indicates that PP and PIB may be miscible in the melt or phase separated at the T_c value studied. This could explain the observed trends of T'_m and G versus compositions, at least for blends not exceeding 20–30% PIB.

Polypropylene–ethylene–propylene random copolymer (PP/EPR) system

The influence of the molecular mass of samples of ethylene–propylene random copolymer (EPR) on the melt rheology, phase structure and properties of PP/EPR blends was investigated by D'Orazio *et al.* [6]. The molecular characteristics of the components are reported in Table 4.3.

The two samples of EPR have practically the same propylene C_3 content and molecular mass distribution, but differ in their molecular mass, as measured by the Mooney viscosity value, with values of 67 and 45. The two samples will be referred to as EPR(HM) and EPR(LM), respectively [6].

The PP and EPR copolymers were mixed in a Werner mixer at 230 °C with a blending time of 3 min. Blends with composition PP/EPR 80/20 (wt/wt) were prepared. After blending, the materials were injection moulded at 230 °C with a mould temperature of 60 °C. The oscillatory shearing flow properties were determined at 200 °C by means of a Rheometrics Mechanical Spectrometer in the plate–plate mode with a constant strain of 10% and an angular frequency ranging between 0.1 and 100 rad/s[1]. The mode and state of dispersion of the minor component was analysed by

(a) (b) (c)

Figure 4.7 Optical micrographs of melt crystallized films of PP/PIB(MM) blends, $T_c = 131$ °C, crossed polarizers: (a) 90/10, (b) 80/20, (c) 60/40 [3].

scanning electron microscopy (SEM) only in the core of the bars in order to eliminate the effect of the mould walls on concentration and shape of the EPR particles [6]. PP/EPR blends exhibit in the whole range of the explored frequency a decrease in viscosity value with increasing frequency, i.e. PP/EPR are pseudoplastic melts. PP/EPR blends show a negative

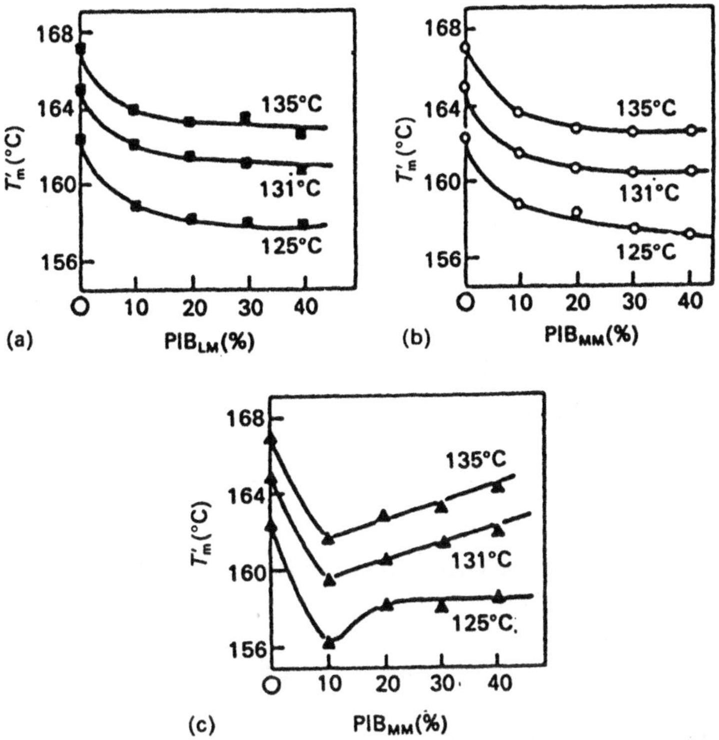

Figure 4.8 Observed optical melting temperature T'_m as a function of the elastomer content at constant crystallization temperature T_c: (a) PP/PIB(LM) blends, (b) PP/PIB(MM) blends, (c) PP/PIB(HM) blends [3].

deviation from the following additive logarithm rule of mixtures that applies at constant temperature and shear rate:

$$\log \eta = \phi_1 \log \eta_1 + \phi_2 \log \eta_2 \qquad (3)$$

where η is the viscosity of the mixture, η_1 and η_2 are the viscosities of the plain components measured at the same temperature, and ϕ_1 and ϕ_2 are their volume fractions.

In oscillatory measurements on polymer melts the frequency (ω) becomes analogous to shear rate ($\dot{\gamma}$). Assuming an approximate equivalence between the complex viscosity (η^*) and apparent viscosity (η_a), the zero shear viscosity (η_0) of both single components and blends was calculated using the Cross–Bueche equation [6]:

$$\eta_0/\eta_a = 1 + (\alpha\dot{\gamma})^m \qquad (4)$$

where α is a parameter that, for polymer melts, is related to the size of the apparent flow unit (the reciprocal of α corresponds to the shear rate at which $\eta_a = \eta_0/2$) and m gives a measure of the shear thinning of the

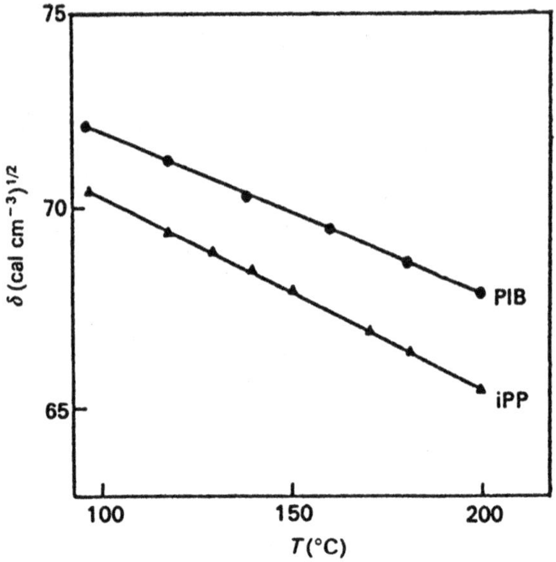

Figure 4.9 Variation of the solubility parameter δ $(cal/cm^3)^{1/2}$ with temperature for PP and PIB [3] (1 cal = 4.18 J).

melt, i.e. a measure of the decrease in viscosity with increasing rate of shear.

From the graph of $1/\eta_a$ versus $\dot{\gamma}^m$ the zero shear viscosity η_0 and α are easily obtained from the reciprocal of the intercept and from the slope, respectively (see Table 4.4). As shown, plain PP at 200 °C has a zero shear viscosity lower than those of pure EPR copolymers. In the case of PP/EPR blends, a large negative deviation from the logarithm additivity rule was observed for η_0. The largest deviation is shown by the blend containing EPR(HM). Thus, the degree of the negative deviation seems to increase with increasing molecular mass of EPR.

The value of α shows that in the blends the transition from Newtonian to pseudoplastic flow starts at a higher frequency than that of plain PP. Moreover, the shift towards higher frequency tends to increase with increasing EPR molecular mass. The analysis of the mode and state of dispersion of EPR, carried out by SEM in the core of transverse microtomed surfaces exposed to boiling xylene vapour to remove EPR, shows that the minor component segregates in almost spherically-shaped domains with dimensions strongly dependent upon the molecular mass of EPR. A finer dispersion is developed in blends containing EPR(LM) (compare Figure 4.10 and 4.11). In such a blend the EPR domains have a diameter ranging between 0.1 and 0.5 μm, whereas in blends containing EPR(HM) the range of particle size is broader (0.1–1.0 μm). The values of the number average particle size (D_n) found for EPR(LM) and EPR(HM) are 0.30 and 0.40 μm, respectively. The dispersion coarseness of EPR copolymers increases with increasing melt

Table 4.3 Molecular characteristics, glass transition (T_g) and observed melting temperature (T'_m) of plain isotactic polypropylene (PP) and ethylene–propylene copolymers (EPR) [6]

Material	*Trade name*	*Melt index* (g/10 min)	*Mooney viscosity* ML (1+4)	C_3 *content* (wt%)	$\bar{M}_w/\bar{M}_n$	T_g(°C)	T'_m(°C)	Code
Isotactic polypropylene	Moplen S30G	1.7	–	–	7.4	−10	166	PP
Ethylene–propylene random copolymer	Dutral COO38	–	67[a]	28	3.5	−40	45	EPR (HM)
	Dutral COO34	–	45[a]	27	3.5	−41	48	EPR (LM)

[a] Measured at 100 °C

Table 4.4 Application of the Cross–Bueche equation: values of zero shear viscosity (η_0), m and characteristic relaxation time (α) for plain PP and EPR copolymers and PP/EPR blends. The shear viscosity calculated assuming log additivity (η' (Pa.s)) is reported for PP/EPR blends [6]

Sample	m	η_0 (Pa.s)	α (s)	η' (Pa.s)
PP	2/3	13071	0.751	–
EPR (HM)	4/7	58227	0.712	–
EPR (LM)	6/11	19973	0.366	–
PP/EPR (HM)	2.3	7285	0.212	176 20
PP/EPR (LM)	4.7	8529	0.528	142 26

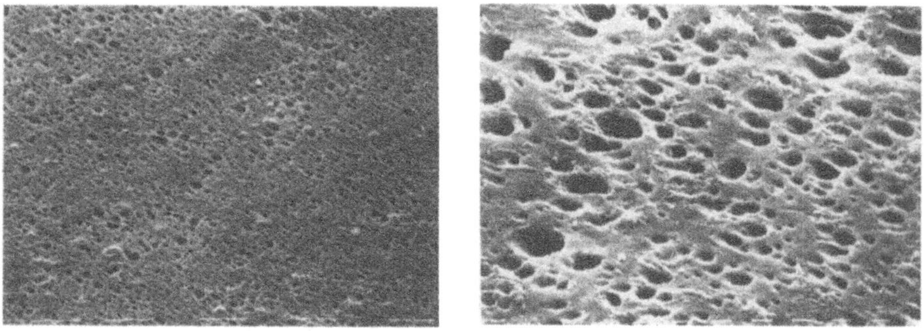

(a) (b)

Figure 4.10 SEM micrographs of smoothed and etched surfaces of PP/EPR(HM): (a) ~ ×780, (b) ~ ×3380 [6].

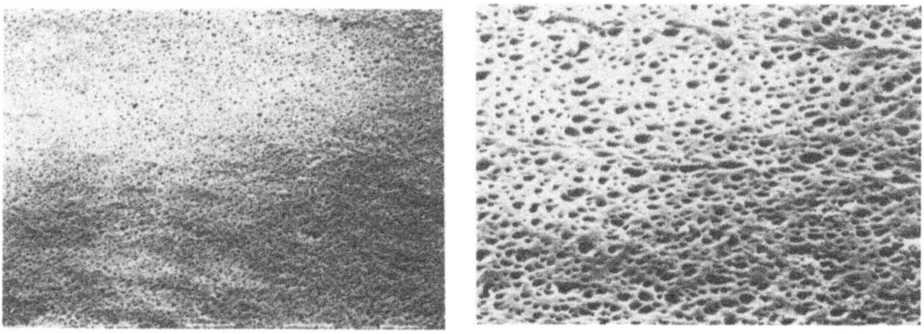

(a) (b)

Figure 4.11 SEM micrographs of smoothed and etched surfaces of PP/EPR(LM) blend: (a) ~ ×780, (b) ~ ×3380 [6].

viscosity, i.e. with increasing phase viscosity ratio (μ), defined as $\mu = \eta_1/\eta_0$, where η_1 is the zero shear viscosity of the disperse phase and η_0 that of the matrix (see Table 4.5) [6].

An opposite trend was observed by Martuscelli and colleagues [7, 8] in polyamide 6/ethylene-co-vinylacetate(PA6/EVA) blends. In this blend it was

Table 4.5 Phase viscosity ratio ($\mu = \eta_1/\eta_0$), range of particle size (D), number-average particle size D_n for PP/EPR blends [6]

Sample	$\mu = \eta_1/\eta_0$	D(μm)	D_n(μm)
PP/EPR (LM)	1.5	0.1–0.5	0.30
PP/EPR (HM)	4.5	0.1–1.0	0.40

established that at constant vinylacetate (VAc) content the average size of the elastomeric particles decreases with an increase in the average molecular mass of the EVA copolymers [7]. Thus, for this blend the average dimension of the dispersed phase decreased with increasing phase viscosity ratio [7].

The type of dependence of the size of the dispersed particles upon the phase viscosity ratio observed in the case of PP/EPR and PA6/EVA blends agrees qualitatively with the predictions of the Rayleigh–Taylor–Tomotika theory [8]. According to this theory, the instability coefficient (q) of a cylindrical thread suspended in a viscous liquid is given by the following expression:

$$q = \frac{\gamma}{2\eta_0 R}(1 - x^2)F(x, \mu) = \frac{\gamma}{2\eta_0 R}\Omega(x, \mu) \tag{5}$$

where γ is the interfacial surface tension, $x = 2\pi R/\lambda$, R is the diameter of the thread, λ is the wavelength of the varicosity of the thread, and μ is the phase viscosity ratio. According to this theory, the plot of D_n versus $\log \mu$ should show a minimum in the vicinity of $\mu = 1$ (Figure 4.12). The different trends in

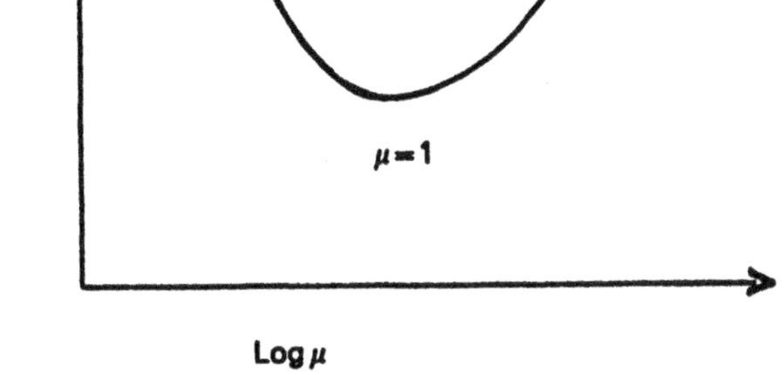

Figure 4.12 Average diameter D_n of dispersed particles as a function of the log μ. Trend as predicted by Tomotika theory [8].

the plot of D_n versus log μ, observed for the two systems mentioned above, were accounted for by assuming that while the data points of PP/EPR blends lie on the right-hand branch of the curve, the opposite occurs for PA6/EVA blends.

The results of the studies on PP/PIB, PP/EPR and PA6/EVA systems showed that the molecular mass of elastomer determines the phase structure of the blend, both in the melt and after the crystallization process. Furthermore, a general correlation was established between the melt rheological parameters and the mode and the state of dispersion of EPR in the melt that is between the values of the number-average particle size (D_n) and the particle size range (D) of the rubbery phase, as determined by SEM on a crystallized sample, and the melt phase viscosity ratio (μ). This correlation was made under the approximation that the crystallization process freezes the melt morphology of the amorphous phase in the melt.

Notched Charpy impact determination showed that for a test temperature higher than the T_g of EPR and PP, the enhancement in PP impact behaviour depended strongly on the molecular characteristics of the dispersed phase (see Figure 4.13). Much better impact properties are shown by the blend containing EPR(HM). At a temperature of 0 °C, the impact value shown by such a blend is about fourteen times as high as that of the neat PP, whereas PP/EPR(LM) blends exhibit an impact value just three times higher than that of unmodified PP. This behaviour was related to the different average particle size of EPR (see Table 4.5). The EPR domains ranging in size between 0.1 and 1.0 μm with an average diameter of 0.4 μm seem to be more effective for PP toughening than particles ranging between

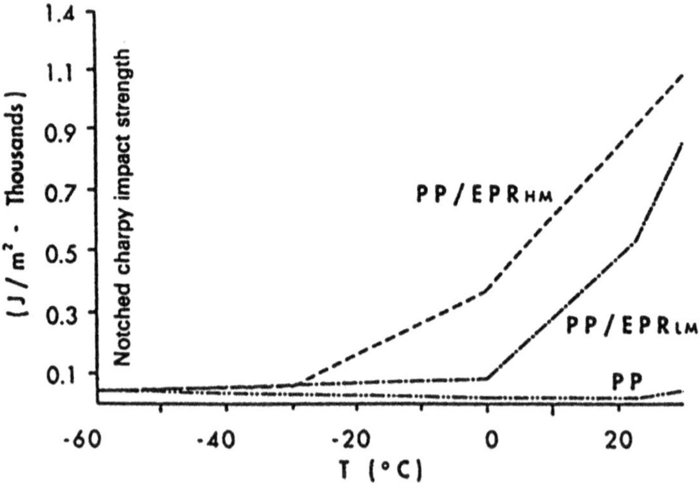

Figure 4.13 Notched Charpy impact strength as a function of temperature for plain PP and PP/EPR blends [5].

0.1 and 0.5 μm. The possibility that in the PP/EPR(HM) blend a more effective interparticle distance is achieved should probably be also considered.

Taking into account that the EPR dispersion becomes coarser with increasing phase viscosity ratio (see Table 4.5) according to the Taylor–Tomotika theory, the parameter responsible for the different degree of dispersity of EPR(HM) and EPR(LM) copolymer turns out to be their different molecular masses. The structure of phases and interphases, developed after complete crystallization of PP, was also studied [9] under controlled crystallized conditions in film samples isothermally crystallized at relatively low ΔT. In the same paper the thermal behaviour and the super-reticular structure of PP phase were also studied. The results may be summarized as follows:

1. The growth rate of PP spherulites is practically independent of the presence of EPR.
2. The apparent melting temperature tends to decrease only when the preliminary crystallization occurred at a low value of ΔT ($T_c = 145$ °C) (see Table 4.6). The equilibrium melting temperature, T_m, of blends is characterized by a depression of 2 to 3 °C. This depression is unaffected by composition and molecular mass of EPR.
3. The lamella thickness (L_c) of PP crystals, as determined by small angle X-rays scattering (SAXS), decreases with increasing EPR content. In the case of 80/20 blends, L_c was reproducibly smaller for samples containing EPR with higher molecular mass (see Table 4.7). The value of the interlamellar thickness (L_a) was found to be larger for blends, this effect also being composition-dependent. It is interesting that even for PP/PIB(LM), a reduction in L_c with increasing PIB content (see Figure 4.14) was found. Thus, it may be concluded that the addition of an elastomer may influence the final intrinsic morphology of the crystallizable component even when rubber and PP are immiscible in the melt.
4. Examination of the crystallized films after complete PP crystallization show that at very low ΔT($T_c = 145$ °C), large amorphous interpherulitic regions develop. It seems that the dimension of such regions increases with EPR content, the effect being more pronounced for blends containing EPR (LM) (Figure 4.15).

   This observation may be accounted for by the smaller dimension of EPR(LM) particles dispersed in the melt, which may be more easily ejected by the crystallizing front.

4.2.2 Influence of molecular mass distribution: PP/EPR systems

The influence of the molecular mass distribution of EPR was studied in the case of an PP/EPR system [11]. The molecular characteristics of the polymers investigated are reported in Table 4.8. In the study, the structure–property relationships of PP/EPR blends prepared with two copolymers (EPR4 and EPR5) having practically the same propylene (C_3) content and mass average molecular mass ($\bar{M}_w$) but differing in their molecular mass distribution ($\bar{M}_w/\bar{M}_n = 4$ and 13.3, respectively) were compared.

The PP and EPR copolymers were mixed in a Banbury mixer at 200 °C with a blending time of 5 min. Blends with composition PP/EPR 80/20 (wt%) were prepared. After blending, the materials were injection moulded at 200 °C with a mould temperature of 60 °C.

From the dependence of the logarithm of the modulus value of the complex viscosity ($|\eta^*|$) upon the logarithm of the investigated frequencies it was concluded that PP/EPR melts are pseudoplastic in the whole range of explored frequencies, i.e. η values decrease with increasing frequency. Furthermore, it was observed that the values of log (η^*) agree with those predicted by Equation 3. This finding disagrees with the results described in the previous paragraph. These results are related to blends containing the same PP but different EPR copolymers. Taking into account that the EPR4 and EPR5 copolymers have a viscosity lower than EPR(LM) and EPR(HM) but close enough to PP viscosity, the contradiction between the results suggests that the melt rheology of PP/EPR blends depends on the molecular structure structure of the rubbery EPR component, which also affects the phase viscosity ratio. The values of the zero shear viscosity η_0 were derived as in the previous paragraph by using the Cross–Bueche equation. The results are summarized in Table 4.9.

It is worth noting that PP/EPR blends, as far as η_0 is concerned, show a positive deviation from the logarithmic additivity rule of Equation 3. Different behaviour was observed for PP/EPR(LM) and EPR(HM) blends. Infact, these blends showed a negative deviation from Equation 3 at 200 °C [6]. The SEM analysis of the mode and the state of EPR dispersion, generated in the PP/EPR injection moulded samples, showed that a layered structure transversal to the mould flow direction, according to the schematic model reported in Figure 4.16, is developed. As shown moving from the border toward the core of the samples, three different layers are found:

1. A skin surface, where almost no dispersed domains of EPR can be observed.
2. An intermediate transition layer, where the concentration of the EPR domains increases going toward the core of the samples with a gradient character. It was found that the thicknesses of the surface skin and of the intermediate transition layer were considerably larger in a blend containing EPR4 than with EPR5 (~40 μm, compared with ~20 μm)
3. A core showing an EPR droplet-like morphology. The range of the diameters of the EPR particles in the core region is reported in Table 4.10, together with the D_n and μ values of the blends.

Table 4.6 Apparent melting temperature (T'_m) and equilibrium melting point (T_m) for PP/EPR blends [9]

Blend	$T_c = 123\,°C$	$T_c = 128\,°C$	$T_c = 133\,°C$ T'_m (°C)	$T_c = 138\,°C$	$T_c = 145\,°C$	T_m(°C)
iPP (100)	162	163	166	169	176	195
iPP/EPR (LM) 90/10	162	162	165	168	173	190
iPP/EPR (LM) 80/20	161	163	165	168	174	191
iPP/EPR (HM) 90/10	161	162	166	169	173	193
iPP/EPR (HM) 80/20	161	162	165	168	174	189

Table 4.7 Lamella thickness (L_c) and interlamellar amorphous thickness (L_a) for PP/EPR blends [9]

	$T_c = 123\,°C$		$T_c = 133\,°C$		$T_c = 132\,°C$		$T_c = 145\,°C$	
	L_c (nm)	L_a (nm)	L_c (nm)	L_a (nm)	L_c (nm)	L_a (nm)	L_c (nm)	L_a (nm)
iPP (100)	8.1	10.8	9.0	12.0	11.1	13.5	13.4	15.7
iPP/EPR (LM) 90/10	6.6	12.3	8.1	14.4	8.7	15.4	10.5	17.1
iPP/EPR (LM) 80/20	5.2	13.2	6.3	15.3	6.9	16.8	9.2	21.6
iPP/EPR (HM) 90/10	6.8	13.1	7.7	14.2	8.7	15.6	10.6	17.4
iPP/EPR (HM) 80/20	4.7	13.5	6.0	15.6	6.6	16.2	8.1	19.0

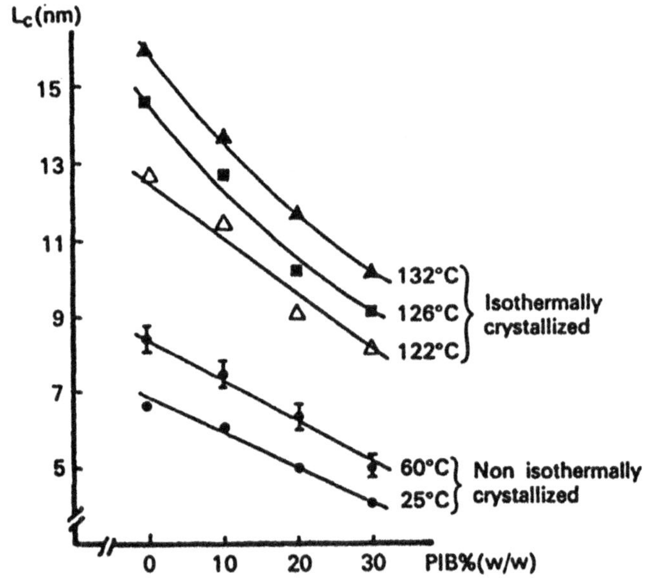

Figure 4.14 Crystalline lamella thickness, L_c, of PP/PIB(LM) blends as a function of elastomer content and temperature, as indicated (1 Å = 0.1 nm) [10].

It can be seen that PP/EPR5 blends exhibit a more coarse distribution of the rubbery particles (D_n is noticeably higher) than those of PP/EPR4 (Figure 4.17).

From the study it is possible to conclude that the molecular mass distribution of EPR represents an important factor in determining the final phase structure of PP/EPR blends. In additional, this affects the final properties of the material, as shown by the impact test results (Table 4.11) From the data in Table 4.11 it emerges that blend samples with EPR4, i.e. with a copolymer of lower molecular mass, show better impact properties when the impact tests are performed at temperatures above the T_g of the EPR.

4.3 INFLUENCE OF CHAIN CONSTITUTION AND MICROSTRUCTURE OF RUBBERY COMPONENT: PP/EPR BLENDS

Greco and colleagues [12, 13] investigated the morphology, structure, development, thermal behaviour and mechanical properties of PP blends containing ethylene (C_2)/propylene (C_3) copolymers with different C_2/C_3 ratios. The EPR samples show a degree of crystallinity that, according to composition, may be related to a solid PP phase (higher C_3 content) or to a polyethylene (PE) phase (higher C_2 content) (see Table 4.12). The influence of copolymer structure and the type of crystallinity on the crystallization, morphology, impact behaviour and drawing process of PP/EPR blends was studied.

Ethylene–propylene copolymers were synthesized in the laboratory by using Ziegler–Natta catalysts. Blends, with a weight ratio PP/EPR = 80/20, of iPP (PPM260, Exxon Chemical Co., with $\bar{M}_w = 440\,000$, $\bar{M}_n = 100\,000$ and $\bar{M}_w/\bar{M}_n = 4.4$) and ethylene–propylene copolymers were prepared by melt mixing in a Brabender-like apparatus at 200 °C. The study was performed on samples crystallized under different conditions:

1. Samples isothermally crystallized from the melt at relatively low undercooling: $T_c = 126$ °C (IC samples)
2. Samples non-isothermally crystallized, obtained by quenching the compression moulded films in cold water (NIC samples).

4.3.1 IC samples

The mode and the state of dispersion of EPR copolymers as a function of C_3 content was studied by SEM using etched (chromic acid) thin sections (Figure 4.18).

With decreasing C_3 content in the EPR copolymer, it was observed that:

1. The average size of PP spherulite decreases slightly and gradually (no apparent spherulites can be seen in Figure 4.18f for the PP/EP16 blend).
2. The spherulites look less and less compact and perfect.
3. More and more voids appear and the interspherulitic boundaries become wider and deeper.
4. The micrograph of PP/EP16 shows a 'Swiss cheese-like' appearance.

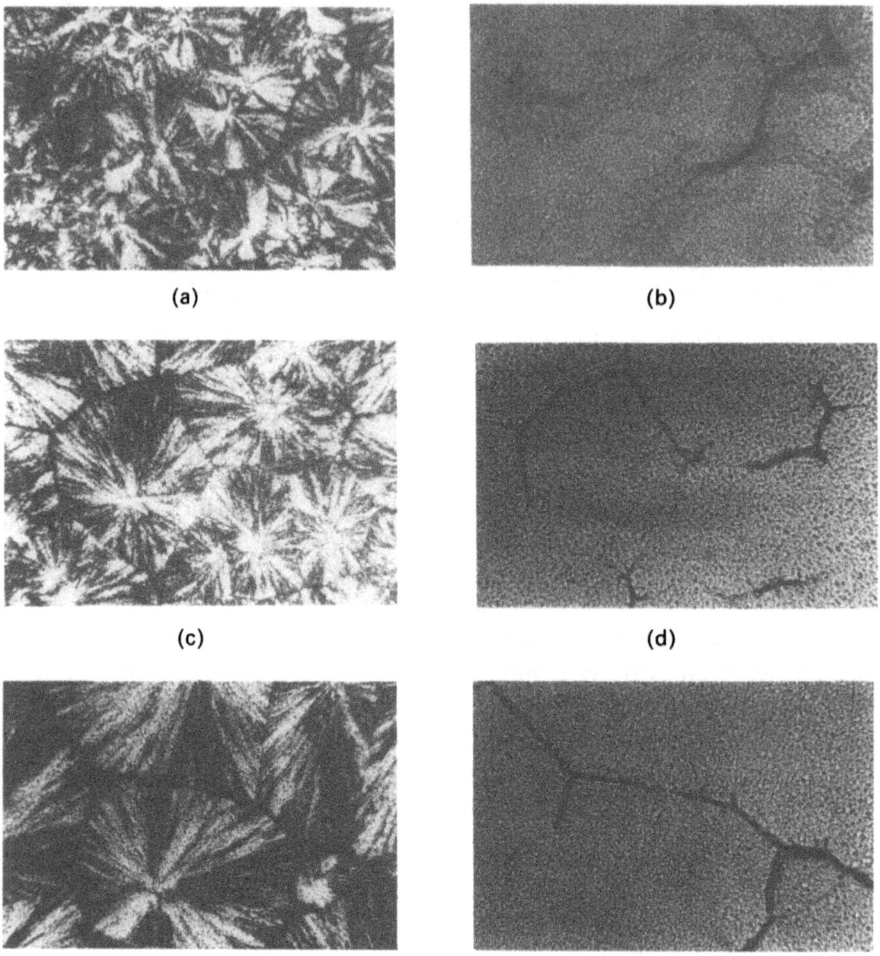

(a) (b) (c) (d) (e) (f)

Figure 4.15 Optical micrographs taken at crossed and parallel polars of: (a–f) PP/EPR(HM) (90/10) and (g–l) PP/EPR(LM) (90/10) [11].

5. The amount of amorphous material removed by etching increases with decreasing C_3 content in EPR.

This last finding was explained by assuming that in the case of EPR copolymers with higher C_3 content the PP blocks inside co-crystallize (partly, at least) with plain PP molecules. Because of this physical 'bonding' between PP and EPR only a relatively small amount of EPR could be etched out from the blends containing EPR copolymers with higher C_3 content.

The values of the overall crystallinity (X_c) as obtained experimentally by DSC agree with X_c^P calculated assuming no interaction beteen PP matrix

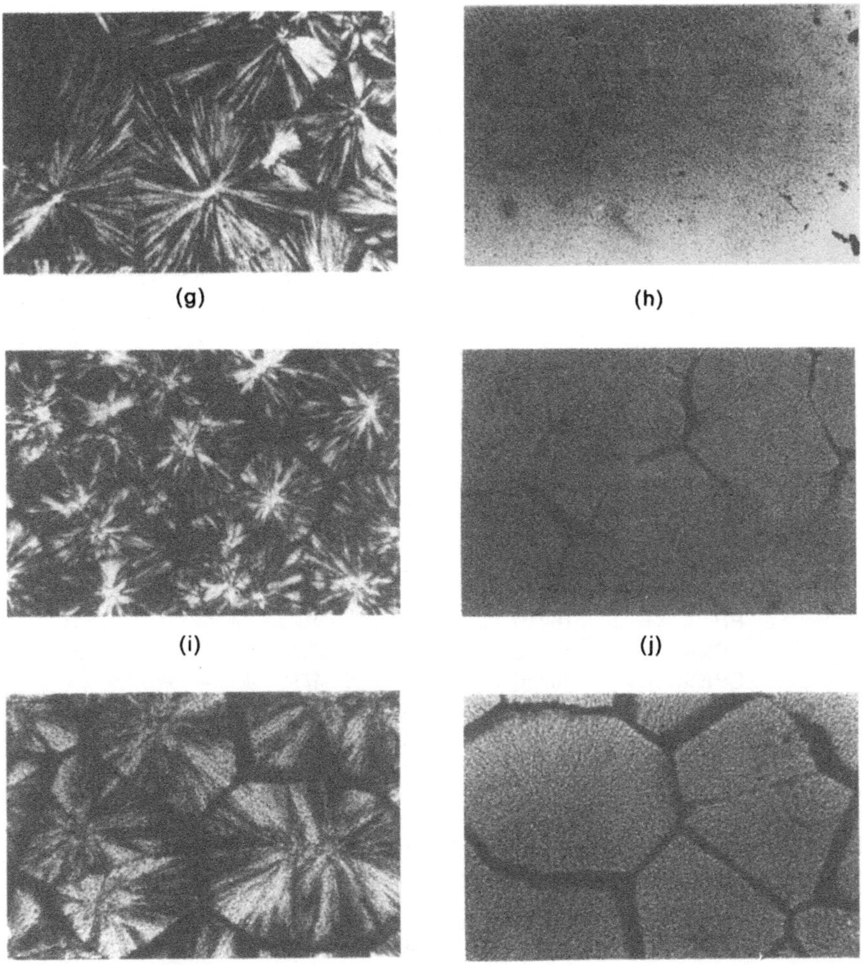

Figure 4.15—*Contd.*

and EPR copolymer domains:

$$X_c^P = W(PP)\,X_c(PP) + W(EPR)\,X_c(EPR) \qquad (6)$$

In Equation 6 W(PP) and W(EPR) are the weight ratios of PP and EPR copolymers, respectively, and X_c(PP) and X_c(EPR) are the crystallinities of plain PP and EPR copolymers crystallized at the same temperature (126 °C) as the blends (see Table 4.13). The change in X_c in the blends with an EPR of lower crystallinity than the PP depended on the molecular structure of the EPR copolymer (see Table 4.13).

As shown in Table 4.13 the melting temperatures of all blends are lower than that of pure PP. This depression can be related both to the creation of

Table 4.8 Molecular characteristics and glass transition temperature (T_g) of neat isotactic polypropylene (PP) and ethylene–propylene copolymers (EPR) [11]

Sample	$\bar{M}_n \times 10^{-3}$ (g/mol)	$\bar{M}_w \times 10^{-3}$ (g/mol)	$\bar{M}_z \times 10^{-3}$ (g/mol)	$\bar{M}_w/\bar{M}_n$	C_3 (wt%)	T_g(°C)
PP	65	484	2782	7.4	–	−10
EPR4	50	200	800	4.0	43	−49
EPR5	15	200	1000	13.3	44	−48

Table 4.9 Application of the Cross–Bueche equation to plain PP and EPR copolymers and to PP/EPR blends: values of zero shear viscosity (η_0), characteristic relaxation time (α) and m; the zero shear viscosity values calculated assuming log additivity rule for PP/EPR blends (η'_0) are also reported [11]

Sample	η_0 (Pa)	α (s)	m	η'_0 (Pa.s)
PP	16410	0.722	2/3	–
EPR4	33102	1.735	4/7	–
EPR5	30750	1.761	4/7	–
PP/EPR4	28736	1.143	4/7	18962
PP/EPR5	26330	1.152	4/7	18677

defects in the PP crystals and to a more general morphological effect. Mechanical tensile tests showed that, due to the low ΔT, all the specimens exhibit a brittle behaviour, with rupture occurring before neck formation (Figure 4.19). The shape of the stress–strain (σ–ε) curves depends on the constitution of EPR and on the degree and type of crystallinity [12].

In the case of blends containing EPR with analogous crystallinity to that of plain PP, both the Young's modulus (E) and the tensile strength (σ_b) increase with C_3 content and then with EPR crystallinity. On the contrary, the elongation at break (ε_b) decreases (Table 4.14). This behaviour is related to the increasing 'physical compatibility' of EPR and PP with the increase of crystalline PP-blocks in EPR chains. The higher crystallinity of EP16 copolymers accounts for the relatively large E value of PP/EP16 in comparison with PP/EP66 blends. The higher elongation at break of PP/EP16 with repsect to all the other blends and to the PP homopolymer was artributed to the microspherulitic texture observed which can give, with respect to the interconnection of the system, an effect analogous to that of a slightly higher crystallization rate, causing a greater PP matrix ductility.

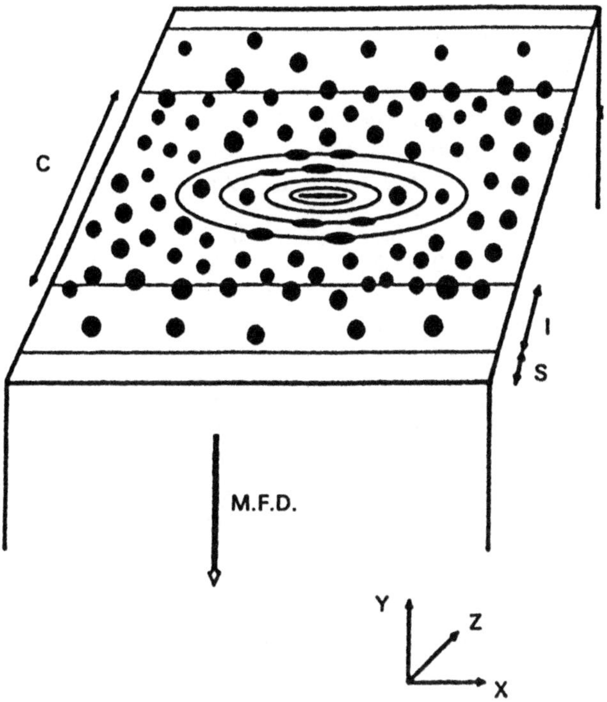

Figure 4.16 Schematic model of layered structure generated in PP/EPR injection moulded samples [11]. (An analogous structuring was reported by J. Karger-Kocsis and I Csikai (1987) *Polymer Engineering and Science* **27**, 241–53.)

4.3.2 NIC Samples

All NIC blend films show a microspherulitic morphology [13]. The phase structure of the NIC-samples was studied mainly by SEM on thin sections etched with boiling *n*-heptane.

The main results are summarized as follows:

1. In the case of PP/EP16 blend a large amount of amorphous material is etched out, unlike plain PP. After extraction, the surface of the film shows distinct holes, indicating that the EP16 copolymer segregates in separate domains with poor adhesion to the matrix.
2. The amount of amorphous material etched out by boiling *n*-heptane decreases progressively in blend films containing EP66, EP74, EP81 and EP88. In the case of PP/ EP88 blend it was not possible to detect any holes after extraction [13]

The above results indicate that increasing C_3 content in the EPR copolymers may cause the system to become more interconnected, due to the fact that the PP blocks along the EPR chains may co-crystallize with the molecules of the plain PP. Thus, EPR macromolecules with higher C_3

Table 4.10 Range of particle size, number average particle diameter (D_n) and phase viscosity ratio (μ) for PP/EPR blends [11]

Sample	*Range of particle size* (μm)	D_n (μm)	μ
PP/EPR4	0.1–0.6	0.35	2.02
PP/EPR5	0.4–1.2	0.80	1.87

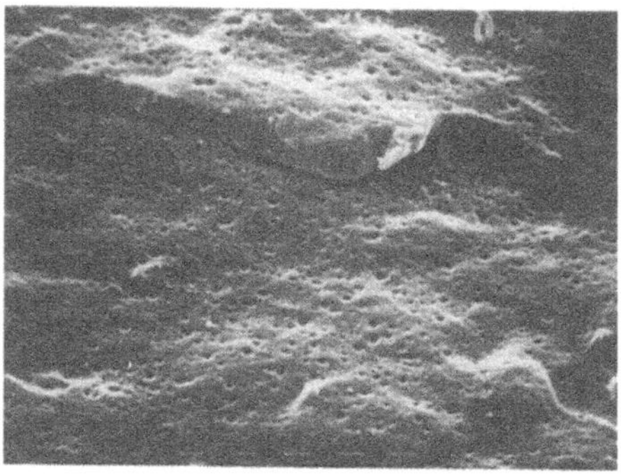

(a)

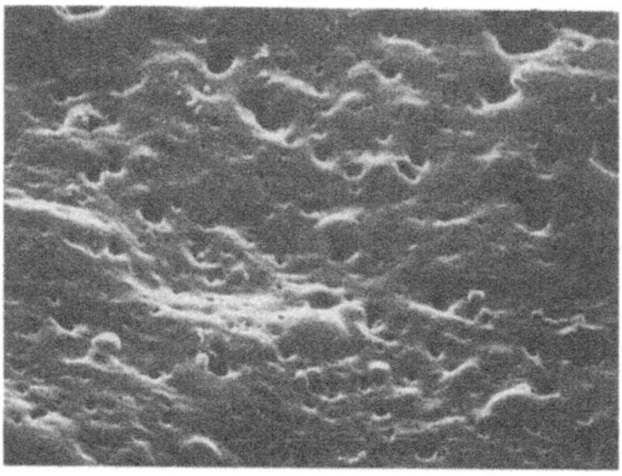

(b)

Figure 4.17 SEM micrographs of fracture surfaces of: (a) PP/EPR4 and (b) PP/EPR5 (80/20) blends at 40 °C, × 5000 [11].

content are physically 'anchored' to the matrix; consequently they cannot be etched out by *n*-heptane.

All the samples except those containing EP16 showed in tensile test a ductile behaviour due to the microspherulitic structure of the PP matrix. The observation of a microspherulitic texture is related to the high rate of cooling, which activates a higher fraction of heterogeneous nuclei. This leads to a high number of spherulites per unit value as well as to crystals with

Table 4.11 Influence of molecular mass distribution of EPR copolymer on the impact behaviour of PP/EPR blends [11]

Test temperature T(°C)	Izod (J/m^2)	
	PP/EPR4	PP/EPR5
+23	742	623
0	144	83
−20	65	56
−40	45	40

Table 4.12 Compositions and characteristics of EPR copolymers [12, 13]

Code	C_3 *content* (wt%)	*Feed ratio of monomers* (C_2/C_3)	$[\eta]$ (dl/g^1) *in decahydronaphthalene at* 135 °C	X_c (%)	T'_m (°C)	T_g (°C)
EP88	92	4/96	1.66	42[a]	143[a]	0–6
EP81	86	14/86	1.83	15[a]	125[a]	−10
EP74	81	22/78	1.67	4[a]	119[a]	−20
EP66	74	32/68	2.08	3[a]	118[a]	−33
EP55	65	44/56	1.65	2[b]	123[b]	−36
EP16	22	75/25	3.73	27[b]	125[b]	−50

[a] PP crystallinity
[b] PE crystallinity

much broader dimensions. The final structure is characterized by a higher concentration of interlamellar tie-molecules among crystallites. This gives a higher level of interaction throughout the blend, yielding a material with more ductile behaviour.

The Young's modulus (E), tensile yield stress (σ_y) and the constant stress at cold-drawing, σ_c (the stress at the plateau of a stress–strain curve) of iPP/EP88, 81, 74, 66 and 16 copolymer blends are shown in Table 4.15. It may be observed that a decrease in the C_3 content in the copolymer lowers all the mechanical parameters of the blends. This is related to the fact that in all of the blends 20% of the PP has been substituted by a less crystalline EPR. Furthermore, on lowering the C_3 content in the copolymers, both the crystallinity content (X_c) and the T_g decrease. This renders the system increasingly soft, and hence E, σ_y and σ_c decrease.

In the case of PP/EP16 blend one should expect higher values of the tensile parameters due to the higher crystallinity of this copolymer. The

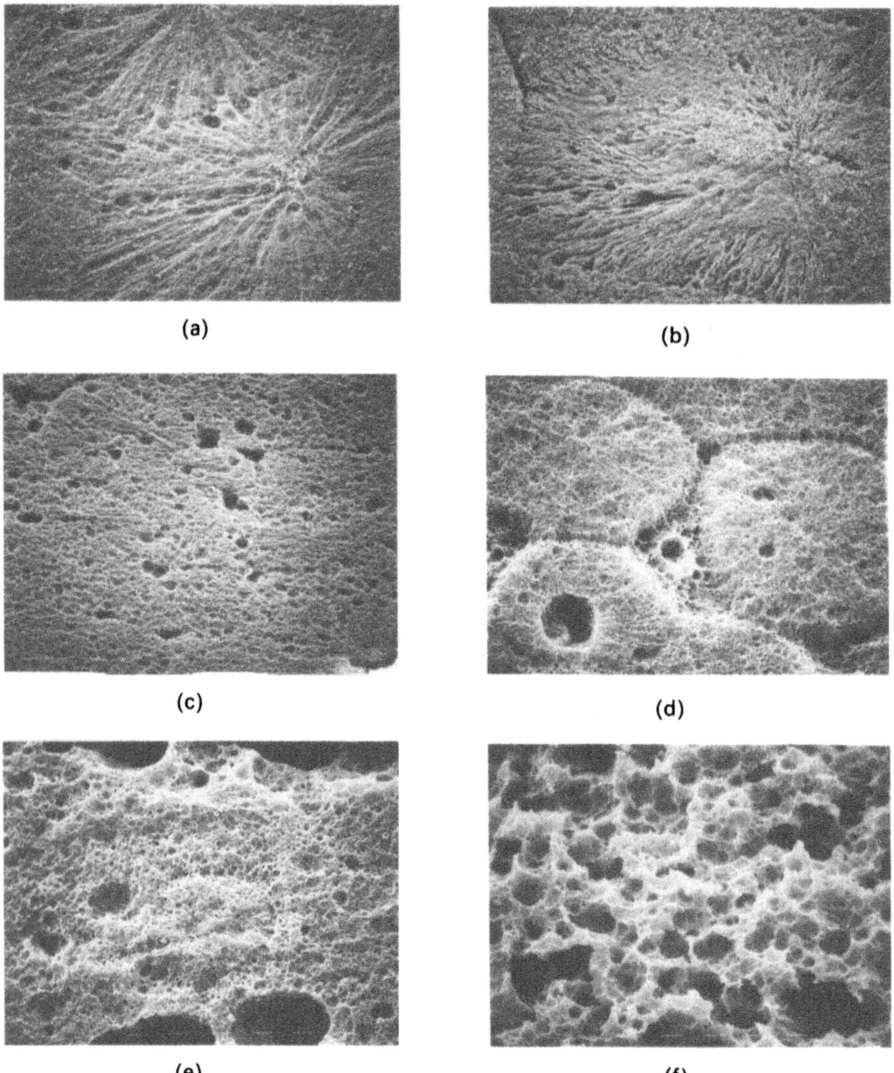

Figure 4.18 SEM micrographs (× 512) of smoothed and etched surfaces of PP and iPP/EP copolymer blends crystallized at 126 °C for 14 h: (a) PP, (b) PP/EP88, (c) PP/EP81, (d) PP/EP74, (e) PP/EP66, (f) PP/EP16 [12].

observation that PP/EP16 blends show values of E and σ_y much lower that those of other blends may be related to a considerably lower compatibility of the two components. Infact, thin films of this blend show, after extraction, holes indicating the presence of large EPR domains with no adhesion to the matrix. This hinders the cold drawing of the samples, causing rupture before crack formation. This is the reason why PP/EPR16 samples are brittle in nature, even if crystallized at high ΔT.

Table 4.13 Index of crystallinity X_c, and X_c^P, melting temperature T'_m and width half-height of melting peaks for PP/EPR copolymers blends ($T_c = 126$ °C, crystallization time = 14 h), as measured by DSC [12]

Sample code	*Index of crystallinity* (%)		*Melting temperature*	*Width at half-heigth of melting peaks*
	X_c	X_c^P	T'_m(°C)	(°C)
iPP	64	64	166	18
iPP/EP88	60	61	163	18
iPP/EP81	58	57	163	17
iPP/EP74	55	54	163	18
iPP/EP66	53	53	163	17
iPP/EP16	59	60	164	18

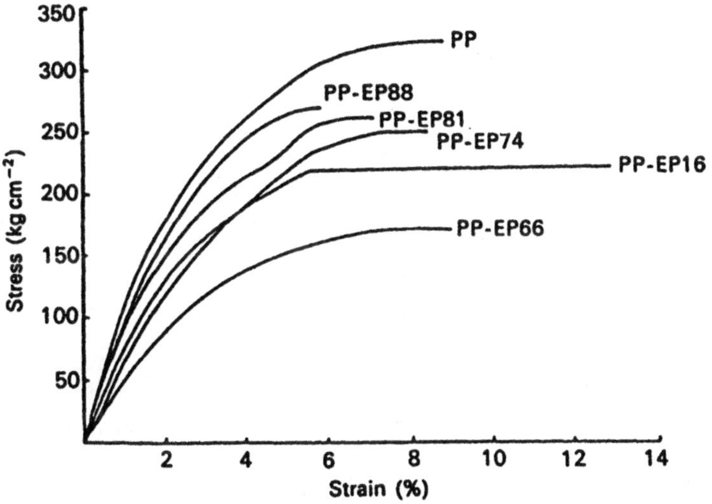

Figure 4.19 Stress–strain curves of PP and PP/EPR copolymer blend specimens, as indicated (1 kg/cm^2 = 0.1 MPa) [12].

The impact strength (R) for the iPP/EP55 and iPP/EP88 blends as a function of temperature is reported in Figure 4.20. The curves refer to different blend compositions [13]. The impact strength of plain PP was very low over the whole range of temperatures investigated. The corresponding fracture surfaces show the typical appearence of a brittle material. For the iPP/EP55 blends (Figure 4.20a) the behaviour at low temperature (up to about 0 °C) is similar to that of PP, but above 0 °C a substantial rise in impact strength is achieved. This rise is accompanied by a stress-whitening phenomenon over the entire fracture surface due to the formation of multicrazes. In contrast, as shown by Figure 4.20b, the iPP/EP88 blends shows no enhancement in R values with increasing temperature. The

Table 4.14 Young's modulus (E) tensile strength (σ_b) and elongation at break (ε_b) of PP and PP/EPR blends crystallized at low undercooling ($T_c = 126\,°C$) [12]

Sample code	E (GPa)	σ_b (MPa)	ε_b(%)
PP	1.1	32	10
PP/EP88	1.0	29	6
PP/EP81	0.8	27	7
PP/EP74	0.7	25	8
PP/EP66	0.5	20	10
PP/EP16	0.7	22	15

Table 4.15 Young's modulus (E) tensile yield stress (σ_y) and the constant stress at cold-drawing (σ_c) of PP and PP/EPR blend [13]

Sample	*Composition* (*PP/EPR;* wt%)	C_3 *content of EPR* (wt%)	E (GPa)	σ_y (MPa)	σ_c (MPa)
iPP	100/0	–	1.14	39	28
iPP/EP88	80/20	92	1.09	36	27
iPP/EP81	80/20	86	1.04	31	24
iPP/EP74	80/20	81	0.98	28	22
iPP/EP66	80/20	74	0.83	26	22
iPP/EP16	80/20	22	0.66	20	

fractured surfaces show no stress-whitening, indicating that fracture occurs without crazing, as in the case of plain PP. The above findings are in agreement with the morphological observations of fracture surfaces. In fact, as shown from the SEM analysis when the C_3 content of the added copolymer is lower, as in the case of EP55, a dispersed phase of spherical rubbery domains can be observed. These rubbery particles, which can be highly adherent to the matrix due to some physical 'interlocking', probably due to co-crystallization with PP blocks, act as sites of stress concentration. Therefore, as soon as the stress around the particles overcomes the yield stress of the matrix, a multiple craze mechanism becomes active and an enhanced toughness is achieved. This effect increases with increasing copolymer content. SEM micrographs demonstrate that in the case of iPP/EP88 blends this type of toughening mechanism is completely absent insofar as no dispersed phase is present.

The thermal behaviour as studied by DSC showed that quenched NIC samples of PP/EPR blends exhibited a higher melting point and narrower melting peaks than plain PP. This enhancement in T'_m was attributed to the 'extraction' of imperfect PP molecules by the EPR phase during the melt mixing, which results in more perfect crystallites in the PP matrix. A similar

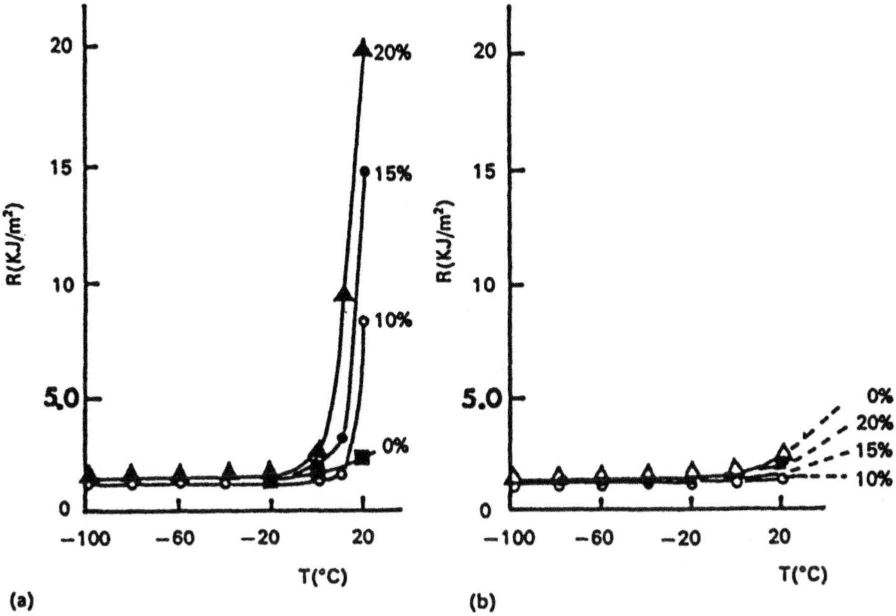

Figure 4.20 Impact strength (R) of pure PP and PP/EPR blends as a function of testing temperature: (a) PP/EP55, (b) iPP/EP88 [13].

effect was reported even in the case of IC samples of PP/EPR blend having different molecular characteristics and preparation conditions [14].

4.4 INFLUENCE OF CRYSTALLIZATION CONDITIONS

The influence of crystallization conditions on the phase structure and properties of PP/elastomer blends was extensively studied for PP/PIB and PP/EPR systems [10, 15, 16]. Samples of PP/(PIB, EPR) blends (composition range in elastomer: 0, 10, 20, 30 wt%) were prepared by melt mixing. Sheets with different thickness (from 0.5 to 1.25 mm) were obtained by pressing the samples between Teflon plates in a press at 200 °C and a pressure of about 20 MPa. After pressing the samples were crystallized in a thermostated bath at temperatures ranging from 20 to 132 °C [10, 15].

4.4.1 PP/PIB system

Sheets samples 0.5 and 1.25 mm thick were:

1. Isothermally crystallized at different undercooling ΔT in the range of T_c 122–132 °C (IC samples).

2. Non-isothermally crystallized by quenching at different temperatures (T_q) in the range 25–90 °C (NIC samples).

The molecular characteristics of samples were: PP $\bar{M}_w = 3.1 \times 10^5$ and $\bar{M}_n = 1.6 \times 10^4$; PIB $\bar{M}_v = 6.6 \times 10^4$ g/mol. As shown in Figure 4.21 the Young's modulus of NIC samples measured at RT decreases with increasing PIB content and with decreasing T_q. The strength behaviour is more complex because, as a function of T_q and percentage PIB, the samples in tensile test behave as ductile (in such a case fibre rupture occurs after neck formation and propagation (cold drawing) or as brittle (breaking occurs before necking). For $T_q = 25$ °C all samples showed a ductile behaviour, whereas for $T_q = 60$ and 90 °C some of the samples were brittle. The fraction of specimens that undergoes a cold drawing at *RT* decreased, for a given T_g with the PIB content, and at constant composition, with the increase T_q (see Table 4.16)

In the case of IC samples the modulus E also decreases with PIB content but, seems to be almost independent of T_c (Figure 4.22). All IC samples underwent a brittle fracture before yield σ_r. The values of σ_r decrease as a function of the PIB content, while for a given composition σ_r seems to

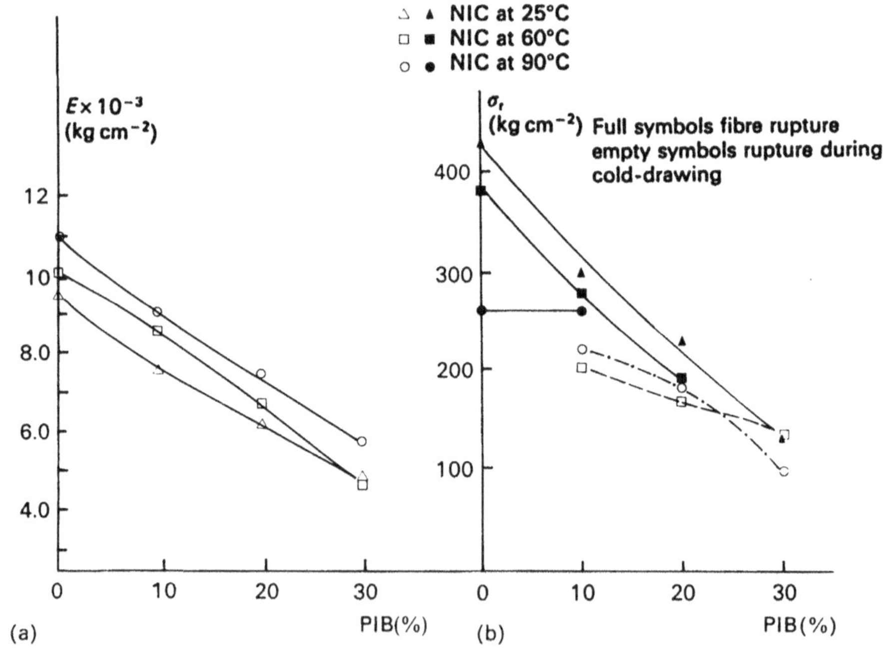

Figure 4.21 Modulus (E) and strength (σ_r) as a function of PIB percentage for specimens non-isothermally crystallized (NIC) at temperatures: 25 °C (triangles), 60 °C (squares) and 90 °C (circles). Open symbols: specimens ruptured during cold-drawing; closed symbols: specimens ruptured after complete fibre formation [10].

Table 4.16 Fraction of PP/PIB specimens fractured as fibres as a function of quenching temperature T_q and PIB content [10]

T_q (°C)	*PIB* (wt%)			
	0	10	20	30
25	1[a]	1	1	1
60	1	0.5	0.2	0
90	1	0.4	0[b]	0

[a] 1, all specimens fractured with fibre formation
[b] 0, all specimens fractured without fibre formation

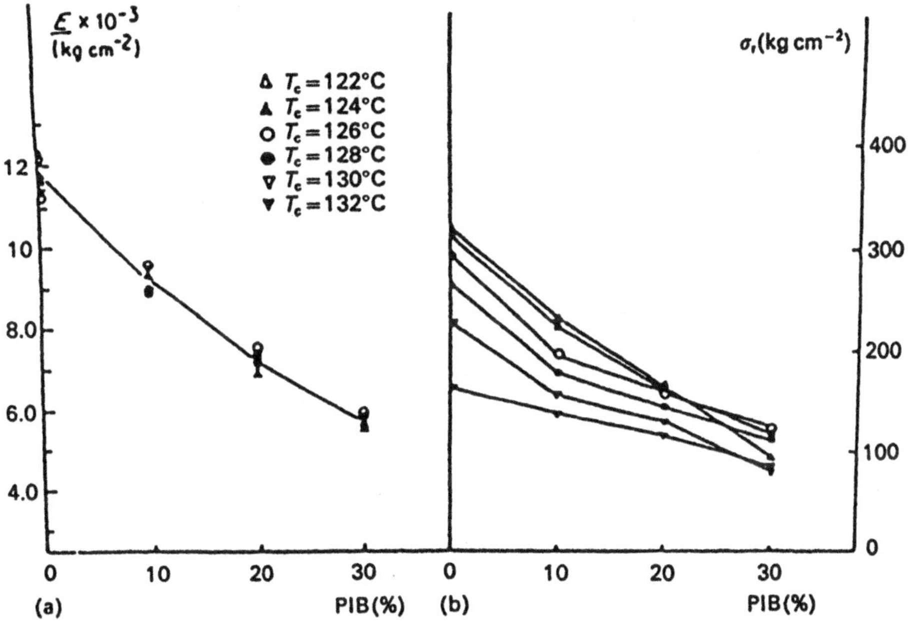

Figure 4.22 Modulus (E) and strength (σ_r) as a function of PIB percentage for specimens isothermally crystallized (IC) at different temperatures as indicated [10].

increase with the decrease of T_c; This effect is especially pronounced at lower PIB content (Figure 4.22).

The phase structure of PP/PIB samples after crystallization was investigated by optical microscopy on thin sections and by wide and small angle X-ray scattering (WAXS and SAXS) [10]. The results are summarized as follows:

1. The IC samples show a double morphology. In the core of the sheets large PP spherulites are observed, while the edges show the presence of 'row-like' structures resulting from a high concentration of heterogeneous

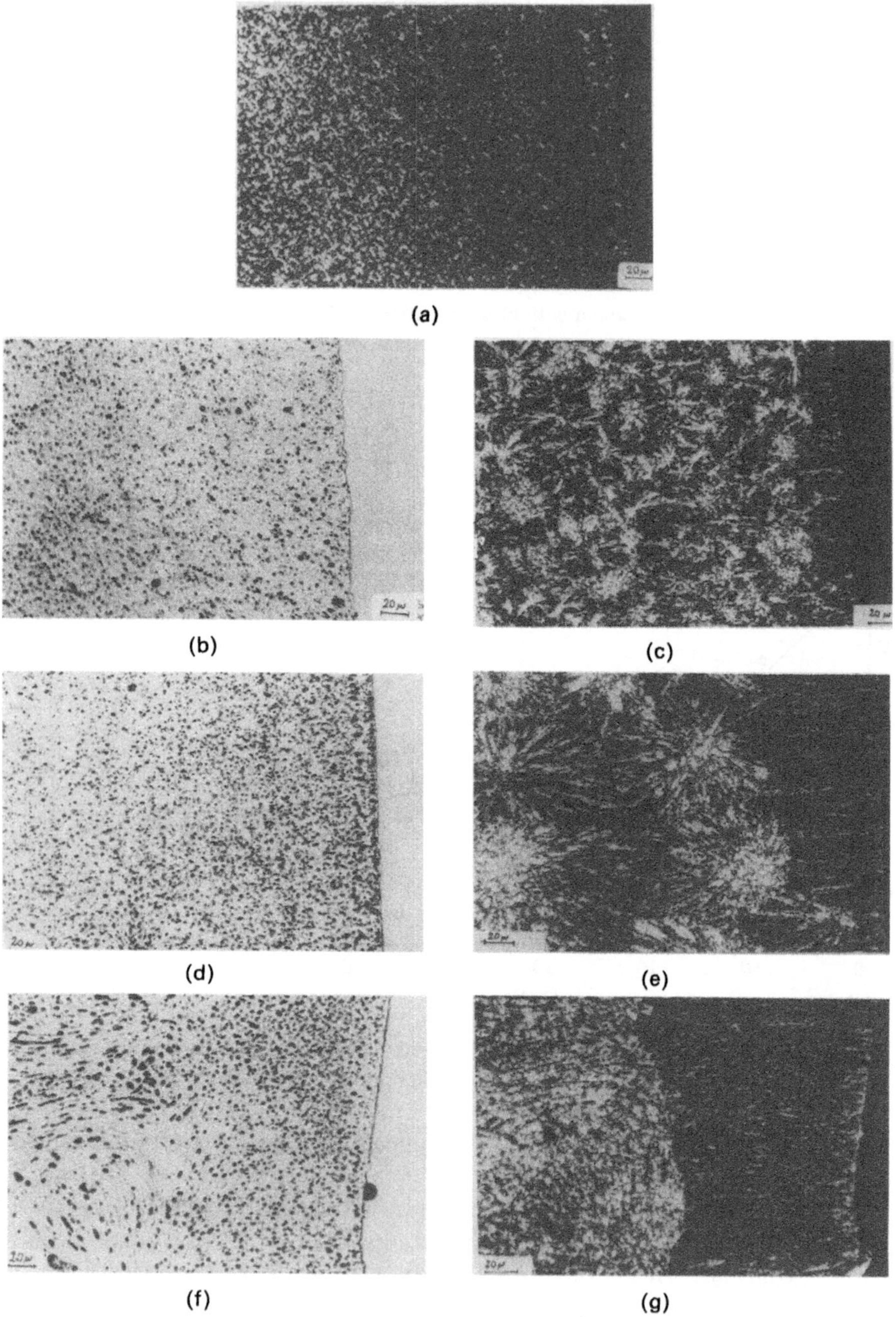

Figure 4.23 Optical micrographs of PP/PIB (80/20) blend obtained at different T_c: (a) $T_c = 25\,°C$ crossed polars (b) $T_c = 90\,°C$ parallel polars, (c) $T_c = 90\,°C$ crossed polars, (d) $T_c = 122\,°C$ parallel polars, (e) $T_c = 122\,°C$ crossed polars, (f) $T_c = 132\,°C$ parallel polars, (g) $T_c = 132\,°C$ crossed polars [9].

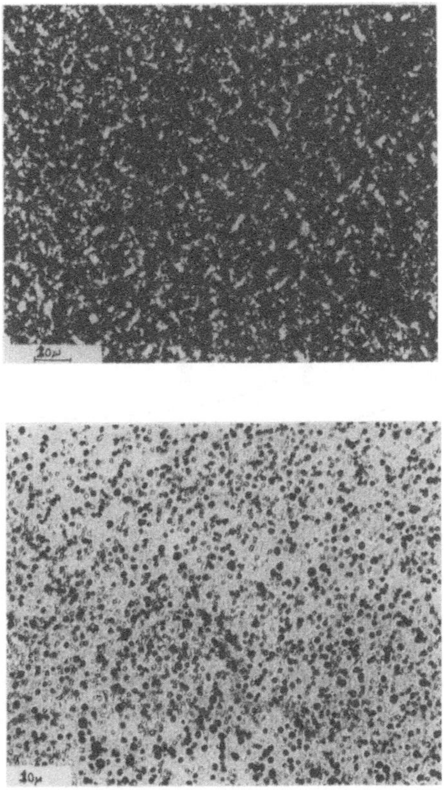

(a)

(b)

Figure 4.24 Optical micrographs of PP/PIB (80/20) blend crystallized at $T_b = 60$ °C: (a) crossed polars and (b) parallel polars [10].

nuclei on the contact-surface of the Teflon sheets (Figure 4.23). The volume fraction of the material that crystallize in such a row-like structure increases with T_c while it seems not to be influenced by PIB content (Figure 4.23).

2. In the case of NIC samples quenched at 25 °C no row structure was detected. All NIC samples are characterized in the core by a microspherulitic structure of PP (see Figures 4.23 and 4.24).
3. The crystallinity of PP phase, for a given T_c, was dependent on PIB content, this effect being more pronounced for IC samples. From this it was concluded that the presence of PIB, especially at low ΔT, influences the perfection of PP crystals and/or reduces the fraction of PP able to crystallize.
4. In the case of IC samples it was found by WAXS analysis that, at constant T_c, the index of the β form of PP is lowered by the addition of PIB. The β index was also dependent upon T_c. In blends crystallized by quenching at 25 and 60 °C no β form was detectable (Figure 4.25).

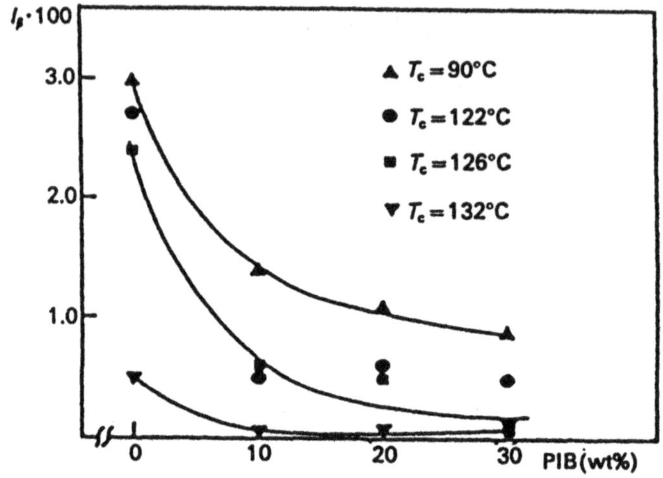

Figure 4.25 β form index, Iβ, of PP/PIB blends as a function of rubber content and temperature, as indicated [10].

5. The desmeared SAXS profiles of all blend samples show defined maxima whose intensities decrease with increasing PIB content (Figure 4.26).

The distance between two adjacent crystalline lamellae of PP, i.e. the long spacing (L), was derived from the maximum position by applying the Bragg equation. Assuming for the spherulite fibrillae a two-phase model, then from

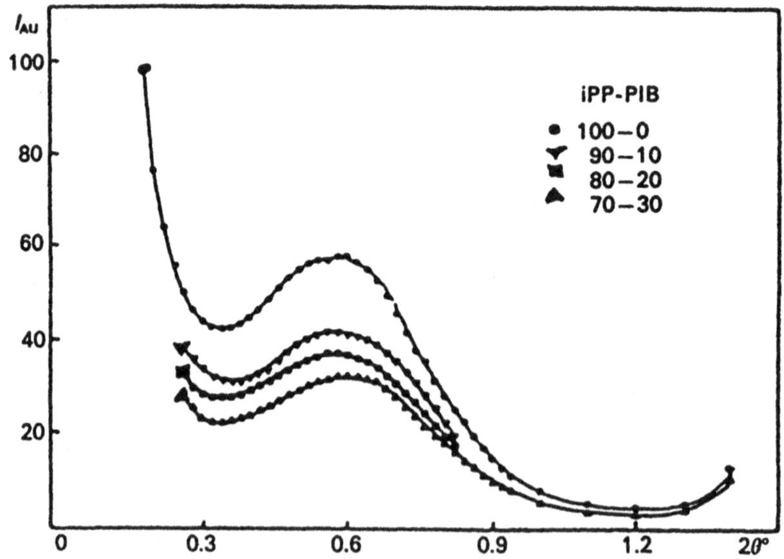

Figure 4.26 Desmeared SAXS profiles of PP/PIB blends non-isothermally crystallized at 60 °C [10].

L the lamellae thickness L_c was calculated as:

$$L_c = \frac{X_c L}{(\rho_c/\rho_a)(1-X_c)+X_c} \tag{7}$$

In Equation 7 X_c is the index of crystallinity of PP phase, ρ_c and ρ_a the densities of crystalline and amorphous PP respectively. For a given T_c and T_q L_c, for both IC and NIC samples, decreases with the PIB content in the blend [10] (see Figure 4.14). This behaviour for L_c is probably accounted for by morphological and kinetic effects mainly caused by the occlusion of spherical domains of PIB in intraspherulitic regions (Figure 4.27). The presence of such particles affects the development of lamellar crystals. It is interesting to point out that, as shown by Figure 4.27, the distribution of PIB particles in intraspherulitic regions is not homogeneous. They are enriched at the spherulitic boundaries due to the ejection phenomenon.

4.4.2 PP/EPR system

The morphology of sheets of PP/EPR blends crystallized at different temperatures, namely 20, 80 and 126 °C, was studied by optical microscopy [15]. A double morphology was observed in all sheets. PP spherulites are

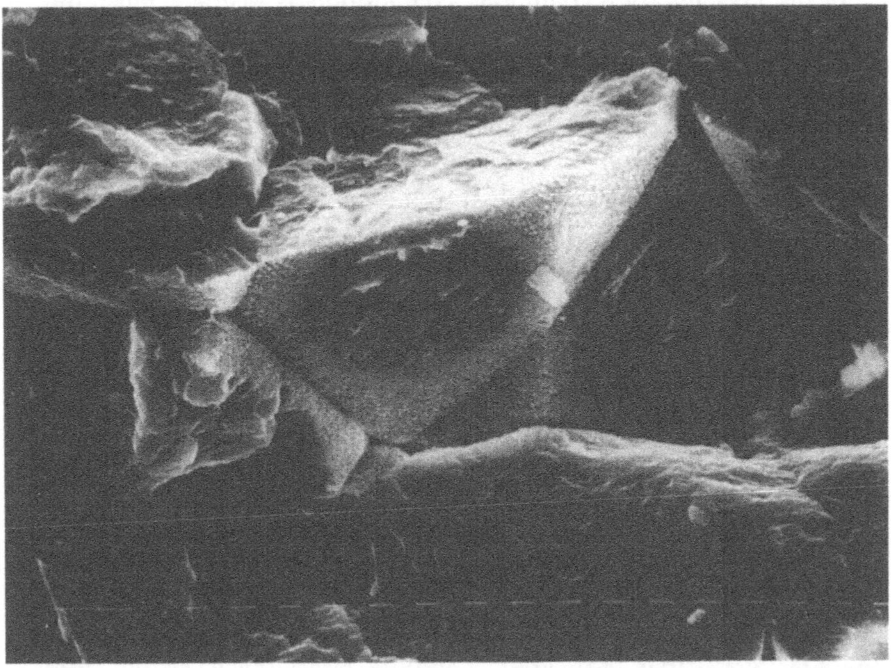

Figure 4.27 SEM micrograph of PP/PIB (90/10) blend crystallized at $T_c = 126$ °C [10].

formed in the centre, whereas a columnar structure is present at the edges. The volume fraction of the material present as columnar structure decreases with decreasing T_c.

The PP average spherulite dimension decreases with increasing undercooling and ranges from approximately 10 μm for $T_c = 20$ °C to 200 μm for $T_c = 126$ °C. The rubber component is dispersed in fine domains of spherical shape in the PP matrix. Their size (ranging approximately from 1 to 4 μm) is dependent on the phase visosity ratio.

At high T_c the rubber particles are much smaller than the PP spherulites and can be rejected or occluded by the growing fronts of the spherulites during crystallization. Therefore, a higher rubber concentration can be found in the interspherulitic regions where some coalescence can occur among the rubber domains. At low T_c this effect tends to disappear since the rubbery particles are suddenly trapped by the fast growing PP spherulites. Since the latter have similar dimensions to the rubbery particles, a more homogeneous dispersion of EPR domains throughout the matrix is obtained. Due to some nucleation effect, discussed in detail elsewhere [17], the average spherulite size of PP for a given T_c, decreases with increasing rubber content. At $T_c = 80$ °C this effect is less evident as the dimensions of PP spherulites are also reduced the decreasing T_c [13].

The mechanical properties of the non-oriented blends were found to be functions of crystallization conditions, composition as well as temperature of drawing (T_d) [15].

At $T_c = 126$ °C with 10% EPR, the blend has a brittle behaviour, with rupture before neck formation for T_d lower than 60 °C (Figure 4.28). On lowering the T_c the same system becomes ductile at all T_d investigated (20–120 °C). On drawing at $T_d = 20$ °C, PP is very brittle for $T_c = 126$ °C, but becomes ductile at $T_c = 80$ °C and $T_c = 20$ °C. If one considers the 80/20

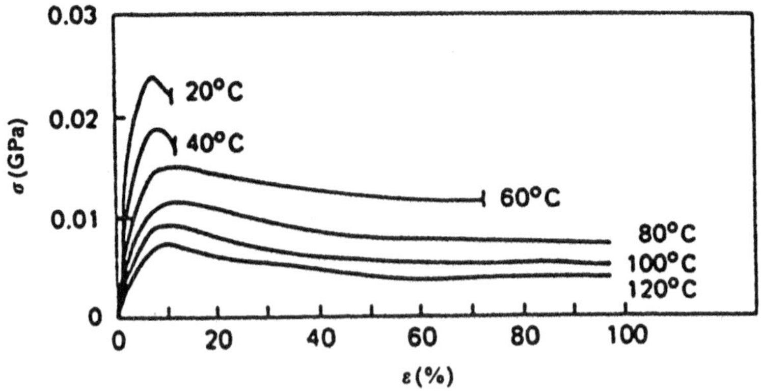

Figure 4.28 Stress–strain (σ-ε) curves of PP/EPR blends containing 10% EPR; crystallization temperature of blends $T_c = 126$ °C, drawing temperature between 20 and 120 °C, crosshead speed of 10 mm/min [15].

blend, the ductility is improved even at 126 °C (Figure 4.29). At $T_c = 80$ °C the system drawn at $T_d = 20$ °C is ductile for all rubber contents investigated (0–20 wt%) (Figure 4.30). The ductility therefore increases with decreasing T_c, with increasing T_d and with increasing rubber content (compare Figures 4.28–4.30)

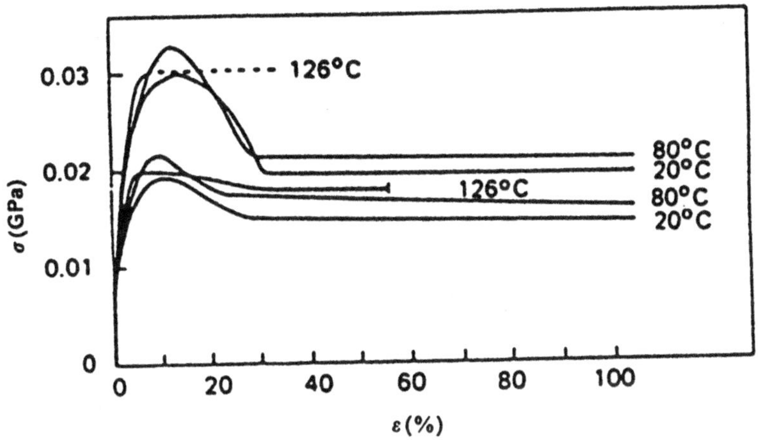

Figure 4.29 Stress–strain curves of PP/EPR blends crystallized at the indicated temperatures and drawn at a temperature of 20 °C [15].

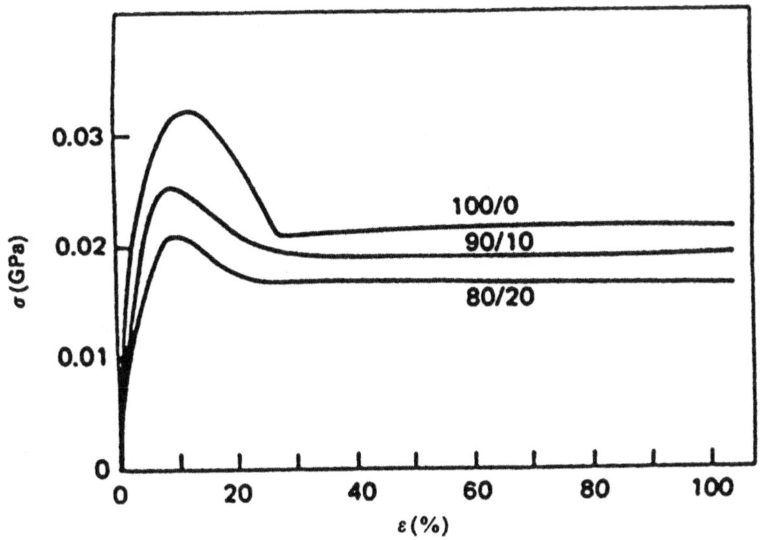

Figure 4.30 Stress–strain curves of PP/EPR blends with different EPR contents; crystallization temperature 80 °C, drawing temperature 20 °C [15].

The influence of T_c on the morphology and properties of PP/elastomer blends was related to the fact that large deformations in semicrystalline polymers depend mainly on the structure of the amorphous layers separating the crystallites and upon the dimension of crystalline lamellae and interlamellar regions and, of course, on their spatial distribution [15].

At low T_c, crystallization occurs very rapidly and the chains are not able to disentangle. The crystallites are formed only locally and the thickness of the lamellae is relatively low. The system is then very interconnected all over the specimen by a high number of tie molecules, i.e. molecules linking the crystallites among them (this is analogous to the entangled network originally existing in the melt). These tie molecules act effectively as local load transducers and yield a ductile mechanical behaviour. In contrast, at high temperatures, which require long times for complete crystallization, the macromolecules in the melt are able to disentangle from each other and to migrate towards the crystalline substrates. In this case only a few tie molecules bridge the relatively thicker formed lamellae. Therefore, when external force is applied the tie chains are unable to carry the load and the lamellae break. Furthermore, at low ΔT molecules with a low tendency to crystallize have the possibility of being rejected into the interlamellar and interspherulitic regions. Under deformation conditions, stress concentrations at such inhomogeneities, present at the boundaries, will promote fracture initiations along the spherulite borders or the intercrystalline regions. The finding that the ductility of samples increase with increasing EPR content was explained by:

1. A decrease of the overall crystallinity of the blend since the rubber is practically a noncrystalline component;
2. A decrease in spherulite dimensions equivalent, perhaps, to a lower crystallization temperature and therefore to a higher degree of interconnection;
3. A 'lubrificating effect' acting on the matrix during the morphological transformation from the spherulitic structure to the fibrillar one during cold drawing, which tends to decrease the internal friction.

Interesting, it was possible to a draw conclusions from the examination of the trend of the curves obtained by plotting E against T_d for a given T_c and blend composition. The E modulus for a given T_c decreases with increasing T_d and also with increasing EPR content. In the modulus versus. T_d diagrams one observes a change in the slope at an interpolated temperature T_i. T_i depends on the initial morphology of the material. For polymer blends crystallized at 20 °C, T_i lies between 40 and 60 °C, whereas T_i is situated between 70 and 80 °C for blends crystallized at 80 °C and increases to approximately 80–100 °C for blends crystallized at 126 °C. An increasing content of EPR seems to slightly lower T_i (Figures 4.31–4.33).

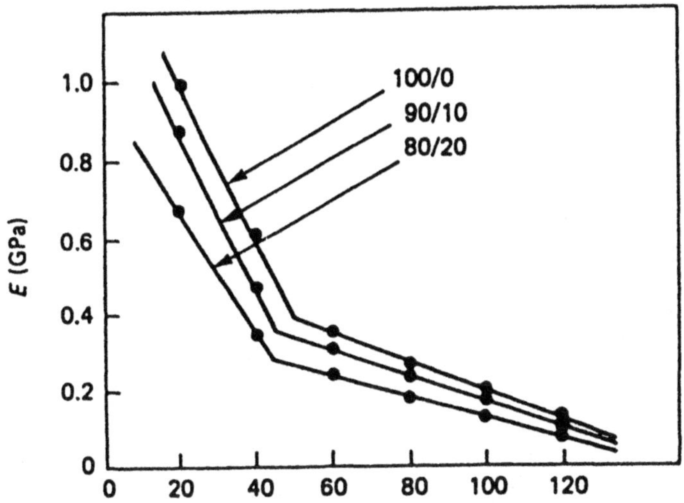

Figure 4.31 *E* modulus as a function of drawing temperature T_d for PP/EPR blends crystallized at 20 °C [15].

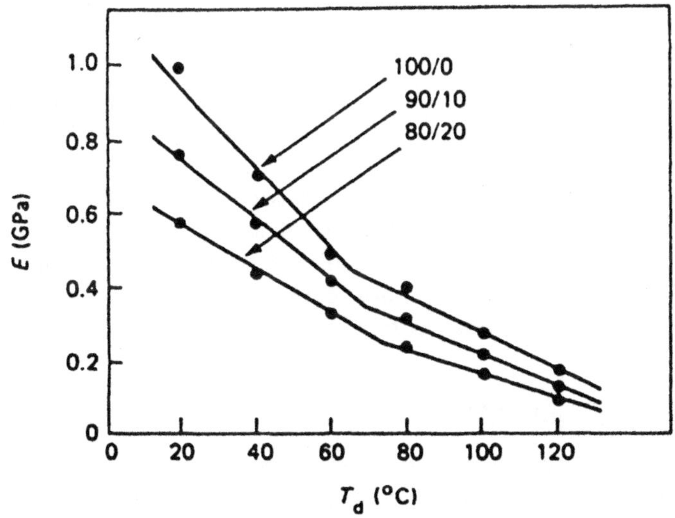

Figure 4.32 *E* modulus as a function of drawing temperature for PP/EPR blends crystallized at 80 °C [15].

At drawing temperatures $T_d < T_i$ all samples became opaque after plastic deformation. At drawing temperatures $T_d > T_i$ transparent samples were obtained. The transition temperature T_i was then related to a change in the drawing mechanism [15]. It was proposed that below T_i the plastic deformation is connected with the formation of microcracks and/or voids, whereas this phenomenon tends to disappear at drawing temperatures higher than T_i [15].

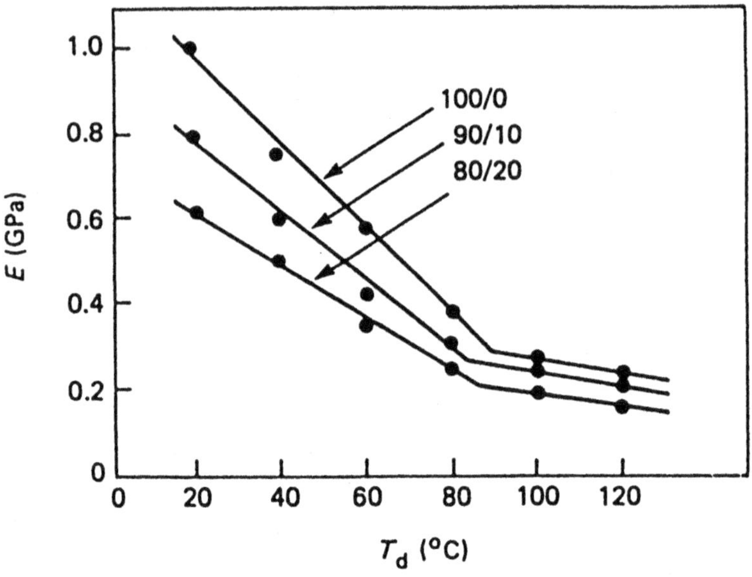

Figure 4.33 *E* modulus as a function of drawing temperature for PP/EPR blends crystallized at 126 °C [15].

It is interesting to note that T_i increases with T_c and seemed almost independent of the blend compositions (compare Figures 4.31, 4.32 and 4.33).

Thus, it can be concluded that the drawing mechanism is strongly dependent on the initial phase structure of the sample (thus on T_c) and on the drawing temperature. The fibre modulus depended strongly on T_d but not on T_c. This suggested that an almost complete cancellation of the initial morphology occurred after fibre formation.

4.5 CONCLUDING REMARKS

It was demonstrated that factors such as molecular mass, molecular mass distribution, molecular structure of components (tacticity and constitution), and crystallization conditions have a great influence on determining the phase structure in the melt and in the condensed state, from which mainly depend the final properties of PP/rubber systems.

It was found that for a given blend composition, mixing procedure and crystallization conditions the mode and the state of dispersion of rubbery domains, in the melt and in the condensed state, after PP crystallization, may be correlated with the melt rheological characteristics of componenets, i.e. with the melt on phase viscosity ratio.

The process of crystallization, especially if it is very rapid (high ΔT), 'freezes' in a first approximation, the melt morphology of the amorphous phase. At the end of crystallization, the material is characterized by the presence of spherulites (larger or smaller according to the density of nucleation and undercooling) that have occluded in intra- and interspherulitic regions of the elastomeric domains. It was demonstrated that the distribution of the rubber particles strongly depends on the ΔT applied affecting the migration of the rubber.

The influence of T_c on the phase structure and then on the properties of PP/rubber blends was interpreted by considering the following. When the blend sample is quenched from the molten state the PP crystallizes at a very high rate. The chain conformations typical of the melt are mostly frozen and only local crystallization can occur. The material will exhibit a microspherulitic texture and crystallites with very thin lamellae. Moreover, the quenching does not allow sufficient time for nuclei alignment and hence no row-like morphology can be expected. Thus a homogeneous and highly interconnected material with a high ductility, able to give a fibre structure after cold drawing, is produced. In fact, the elevated number of tie molecules, coupled with the intrinsic weakness of the thin crystallites, favours the crystal 'fragmentation' necessary to their reorientation in the necking zone.

If the samples are crystallized at low ΔT then the crystallization rate and the system tend to approach equilibrium conditions of crystallization. In such a case only a few nuclei are activated and a macrospherulitic texture is attained with larger crystallites.

The brittle behaviour is due to the low number of tie molecules, able to carry the load, linking spherulites and crystallites, as well as to the high resistance to fragmentation given by the thicker and more perfect crystals. The rubbery particles are partly rejected during the crystallization by the advancing front of the growing spherulites. The material is characterized then by a non-homogeneous distribution of particles.

For a given crystallization process the final properties of PP/rubber blends will only be partly controlled by the mode and state of dispersion of the rubbery component at the service temperature. In fact, the rubbery component interferes with both the primary and secondary nucleation processes and with kinetic and thermodynamic factors of the crystallization of PP, inducing deep modification in its final phase structure, i.e. in the texture, dimension and size distribution of the spherulites, in the inner structure of the spherulites (lamellar and interlamellar thickness), in the physical structure of interspherulitic boundary regions and amorphous interlamellar regions and in the nature of the molecular interconnections active between these structural elements (number and type of tie molecules).

The physical behaviour, as well as the final use properties, of PP/elastomer pairs will be the result of a complex combination of several factors, not easily rationalized.

Nevertheless, an optimization of the mode and state of dispersion of the rubbery particles and of the crystalline texture can be accomplished by selecting the two components according to their molecular mass, molecular mass distribution, constitution and tacticity and by choosing crystallization conditions suitable to produce a stable overall phase structure able to impart desired properties to the material.

4.6 ACKNOWLEDGEMENT

This work was partly supported by PF, Chimica Fine, CNR Italy.

4.7 APPENDIX: SYMBOLS AND ABBREVIATIONS

C_2	Ethylene content of the polymer
C_3	Propylene content of the polymer
D_n	Number-average particle diameter
E	Young's or E modulus
E_1	Activation energy term containing the contribution of the rubber rejection
EPR	Ethylene/propylene random copolymer
EPDM	Ethylene/propylene/diene terpolymer (from ethylene/propylene/diene monomers)
EVA	Ethylene/vinylacetate copolymer
G	Spherulite growth rate
HM	High mean molecular mass
iPP	Isotactic polypropylene
IC	Isothermally crystallized sample
Iβ	Index of the β-form crystallinity
L	Long distance or spacing between lamellae
L_a	Interlamellar thickness (amorphous interlayer)
L_c	Crystalline lamella thickness
L_M	Low mean molecular mass
MFD	Mould filling direction
MM	Medium mean molecular mass
M_n	Number-average molecular mass
M_v	Viscosity-average molecular mass
M_w	Mass (weight)-average molecular mass
M_z	z-average molecular mass
NIC	Non-isothermally crystallized sample
PA6	Polyamide-6
PIB	Polyisobutylene
PP	Polypropylene (homopolymer)
q	Instability coefficient
R	Impact strength
RT	Room temperature

SAXS	Small-angle X-ray scattering
SEM	Scanning electron microscopy
T_c	Crystallization temperature
T_d	Drawing temperature
T_g	Glass transition temperature
T_i	Interpolated temperature
T_m	Equilibrium melting temperature
T'_m	Apparent melting temperature
T_q	Quenching temperature
W	Mass (weight) faction
WAXS	Wide-angle X-ray scattering
X_c	Crystallinity
Y	Shear rate
α	Parameter considering the apparent flow unit in a melt
ΔT	Undercooling or supercooling
γ	Interfacial surface tension
δ	Solubility parameter
ε_b	Elongation at break
η	Melt viscosity
η^*	Complex viscosity
η_a	Apparent viscosity
η_0	Viscosity at zero shear rate
μ	Phase viscosity ratio
ρ	Density
ρ_a	Density of the amorphous phase
ρ_c	Density of the crystalline phase
σ_b	Ultimate tensile strength
σ_c	Constant stress at cold-drawing
σ_r	Breaking stress
σ_y	Tensile yield strength
ϕ	Volume fraction

4.8 REFERENCES

1. Martuscelli, E., Palumbo, R. and Kryszewski, M. (eds) (1980) *Polymer Blends: Processing, Morphology and Properties*, Vol. I, Plenum Press, New York.
2. Martuscelli, E., Palumbo, R. and Kryszewski, M. (eds) (1984) *Polymer Blends: Processing, Morphology and Properties*, Vol. II, Plenum Press, New York.
3. Martuscelli, E., Silvestre, C. and Bianchi, L. (1983) *Polymer*, **24**, 1458.
4. Bartczak, Z., Galeski, A. and Martuscelli, E. (1984) *Polymer Engineering and Science*, **24**, 1155.
5. Martuscelli, E. (1985) Relationships between morphology, structure, composition and properties in isotactic polypropylene based blends, in *Polymer Blends and Mixtures* (eds D. J. Walsh, J. S. Higgins and A. Maconnachie) Martinus Nijhoff, Dordrecht.

6. D'Orazio, L., Mancarella, C., Martuscelli, E. and Polato, F. (1991) *Polymer*, **32**, 1186–94.
7. Addonizio, M. L., D'Orazio, L., Mancarella, C. and Martuscelli, E. (1989) *Journal of Materials Science*, **24**, 2939–50.
8. Martuscelli, E. (1990) Rubber modification of polymers: phase structure, crystallization, processing and properties, in *Thermoplastic Elastomers from Rubber–Plastic Blends*, (eds S. K. De and A. K. Bhowmick) Ellis Norwood, London, pp. 28–70.
9. D'Orazio, L., Mancarella, C., Martuscelli, E. and Sticotti, G. (1991) *Journal of Materials Science*, **26**, 4033–47.
10. Bianchi, L., Cimmino, S., Forte, A. *et al.* (1985) *Journal of Materials Science*, **20**, 895–905.
11. D'Orazio, L., Mancarella, C., Martuscelli, E. and Sticotti, G. (1993) *Polymer* **34**, 3671–81.
12. Greco, R., Martuscelli, E., Ragosta, G. and Yin Jinghua (1988) *Journal of Materials Science*, **23**, 4307–14.
13. Greco, R., Mancarella, C., Martuscelli, E. *et al.* (1987) *Polymer*, **28**, 1929.
14. Martuscelli, E., Silvestre, C. and Abate, G. (1982) *Polymer*, **23**, 229.
15. Coppola, F., Greco, R., Martuscelli, E. *et al.* (1987) *Polymer*, **28**, 47.
16. Kammer, H. W., Kummerlöwe, C., Greco, R. *et al.* (1988) *Polymer*, **29**, 963.
17. Bartczak, Z., Galeski, A., Martuscelli, E. and Janik, H. (1985) *Polymer*, **26**, 1843.

5

Oriental drawing of polypropylene and its blends

S. V. Vlassov and V. N. Kuleznev

5.1 INTRODUCTION

Oriental drawing is a process of great interest for many polymers, because it is a powerful method for improving the mechanical properties. Polypropylene (PP) is one of the most suitable polymers for drawing. It can be drawn to high elongations in acceptable intervals of technological parameters. Fibres of PP can be so highly oriented that it was formerly referred to as 'steel-like'.

In recent years, drawing technology has reached a very high level. Under given conditions, it has become possible to create a crystalline structure composed of extended chain crystals, giving the possibility of producing materials with high tensile strength as well as high thermal stability. Nevertheless, the present technology of orientational drawing is based mainly on traditional methods. These methods allow the achievement of high orientations in polymer films with folded chain structure.

In this chapter the theory and practice of orientational drawing of PP and PP-based blends are summarized.

5.2 ORIENTATIONAL DRAWING OF PP

5.2.1 Uniaxial orientation

The orientational drawing process of PP films is described in detail in the literature [1–3]. The problems associated with uniaxial orientation of isotactic PP (iPP) in films, sheets, fibres and tapes are treated comprehensively.

Polypropylene: Structure, blends and composites. Edited by J. Karger-Kocsis.
Published in 1995 by Chapman & Hall, London. ISBN 0 412 58430 1

The characteristic feature of the orientation of iPP is that its deformation, up to its melting temperature T_m occurs through necking formation. Drawing at temperatures above T_m is followed by a decreasing of the specimen thickness which is proportional to the elongation ratio λ.

In Figure 5.1 the typical morphology of a PP-film in the early stage of deformation is shown [4]. Between the area of isotropic, nondrawn material (1) and oriented material (2) a small area can be seen (3) where the polymer was deformed at very high speed and with intensive heat evolution. The existence of area 3 follows from the obvious fact that the polymer in area 1 is still not oriented, whereas the polymer in area 2 is already fully oriented. The size of area 3 depends on the deformation conditions and on the properties of PP, and varies in the range 0.1–3 mm. It is important to emphasize that at the real applied speed of uniaxial PP film drawing, the relative speed of polymer deformation in this thin area may reach several hundreds or thousands of percentage points per minute. This extremely high speed of relative elongation in area 3 does not permit disorientation of macromolecules to occur during drawing. As a consequence, no significant plastic deformation occurs up to temperatures very close to T_m. The ultrahigh deformation speed in area 3 leads to sudden fibrillization of macromolecules with microcells between the fibrils. When the deformation is released 'free', i.e. without clamps fastening the film in the transverse direction, the polymer contracts (region 4 in Figure 5.1) and the film becomes narrower during uniaxial drawing. The wider the film, the more

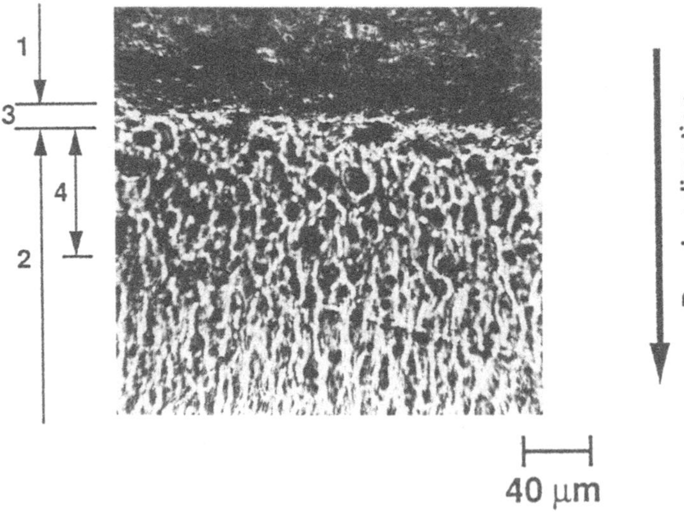

Figure 5.1 The morphology of uniaxially oriented PP film: 1, nonoriented; 2, oriented; 3, region of ultrahigh-speed orientation; 4, region of transverse shrinkage development. Arrow, direction of drawing.

difficult and retarded is this process of narrowing (shrinkage). When shrinkage occurs, the microcells disappear (area 4) and the polymer can not be drawn in the transversal direction. This phenomenon is of great practical significance: samples with lower shrinkage have better ability for transversal drawing during biaxial orientation.

Orientational drawing of PP films leads to substantial changes in their mechanical properties. Figure 5.2 show the dependence of the tensile strength on the drawing ratio, λ, in both orientation and transverse directions, as well as the dependence of the light birefringence (Δn) on λ. The feature of the interrelationship between σ and λ changes insignificantly with the drawing temperature. Beginning from a λ value of 4.5–5 the whole isotropic film transforms to neck. This λ value is called the degree of natural drawing ratio. Rather important is the value of limiting drawing ratio, λ_{max}, which is different for films, fibres and sheets of PP. This value also depends on the method of drawing: formation of oriented film or fibre from a

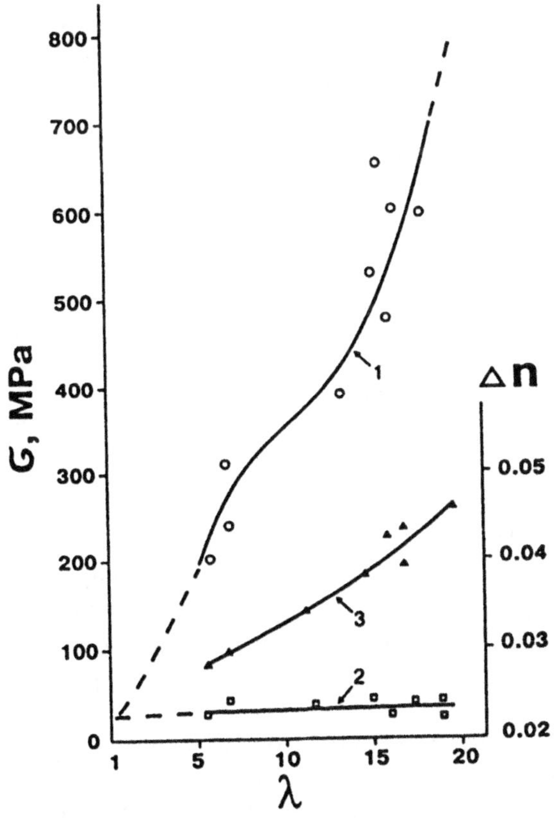

Figure 5.2 (1, 2) Tensile strength σ, (3) light birefringence Δn, versus drawing ratio λ of PP film. Tensile strength: 1 in orientation direction, 2 in transverse direction.

solution or from a melt, and so on. For example, when PP fibres are drawn in conditions favourable to macromolecule orientation [2], λ_{max} may reach values between 10 and 20 or even greater than 50. At the same time, in the orientation process of regular PP films λ_{max} is no greater than 10–12. Nevertheless, the degree of molecular orientation f in the crystalline phase of the polymer may reach a value of 0.96–0.96. In the process of PP orientation, there is a rather small increase in crystallinity (Table 5.1). Noticeable increase in crystallinity takes place after 'superdrawing', i.e. when $\lambda > 30$. With increasing λ both the tensile strength, σ, and E modulus also increase in the superdrawing stage. An analogous sharp increase of σ can be observed close to the limiting drawing ratios in the case of hard-chain polymers. So, during stretching of poly-p-phenyleneterephthalamide (PPhTA) fibres, σ increases steadily up to 0.2–0.25 GPa. At the same time, the Hermans orientation factor (f) also increases up to 0.96. When the f value reaches 0.99, σ becomes 1.5 GPa [6]. It is difficult to reach high f values, especially for wide films, because they fail at early stages. Preparation of films with very high f value is also undesirable for another reason: such films cannot be oriented in the transversal direction because they fail at the very beginning of stretching. On the other hand, deformability in the transversal direction is necessary in order to obtain biaxially oriented films with high strength in both directions.

Table 5.1 Properties of drawn PP films [5]

Drawing ratio, λ	*Tensile strength* (GPa)	*Birefringence* ($\Delta n \times 10^{-3}$)	*Crystallinity* (%)
30	0.54–0.88	19	77.4–78.9
60	1.04–1.24	23	80.4–81.3
80	1.15–1.43	25	81.1–83.3
100	1.35–1.56	27	82.6–84.4

5.2.2 Biaxial orientation

In order to prepare PP films that are 'strong' enough in both directions, biaxial orientational drawing should be applied. Such films can be prepared using two different procedures [7, 8]: (i) simultaneous film drawing in two perpendicular directions, and (ii) stepwise drawing in one direction and then in the other.

In simultaneous biaxial orientation the mechanical properties of the film improve with increasing λ in both directions. The morphology of such PP films looks like a network of interconnected fibrils oriented equally in both directions. Actually, the orientation of fibrils also depends on the difference in drawing ratios in both directions: only at equal values of the drawing ratio do the fibrils have similar structure and are oriented in equal proportions in both directions.

In the process of stepwise biaxial orientation the original film is uniaxially oriented with morphology as shown in region 2 of Figure 5.1. Transversal drawing of such a films leads to a new structure formation with perpendicular orientation of fibrils [4]. Owing to transversal drawing, a new neck forms. The boundary between the neck and the other part of the sample is smooth and wide in the case of drawing in the transversal direction. During stretching in the perpendicular direction the film must be fastened at the edges to prevent shrinkage, otherwise orientation of macromolecules in the first direction will not be preserved. In this process fibrils (region 2 in Figure 5.1) change their positions – one part of each fibril keeps the initial position while another part turns to an angle of 90° In this way a specific morphology forms, in which interperpendicular parts of fibrils are 'knitted' together in one structure. Such a framework of fibrils provides a high level of mechanical properties in both directions. Quite naturally the higher is λ in the transverse direction, the greater is the tensile strength in this direction (compared to the originally nonoriented film). The high value of the tensile strength in the first direction remains almost constant during the second stretching.

5.3 ORIENTATIONAL DRAWING OF PP-BASED HETEROPHASE BLENDS

Certain specific features are characteristic of orientational drawing of PP-based blends. Many works have been published on the properties of PP-based blends. Far fewer works, however, have concentrated on the problems of uniaxial drawing, and fewer still on biaxial orientational drawing of such compositions. The films composed of blends may be divided into several groups, according to the modulus ratio of their components and yield stress values of the blends. These values were determined at a specific temperature and deformation rate of the film drawing process [9]. The next grouping of PP-based blends is based on the modulus ratio of the blend components.

1. The modulus of the dispersed phase is much lower compared to that of the matrix polymer (PP). In this case the PP phase is oriented. The dispersed polymer remains nonoriented although its particles become partially elongated. During orientational drawing such films are deformed homogeneously. No matrix separation from the dispersed particles occurs under stretching. The film surface remains smooth and its appearance does not change. The properties of the oriented film are governed mainly by those of the PP with certain influence of the soft dispersed particles.

2. The modulus of the PP is close to that of the dispersed phase. In this case drawing can take place in both polymeric phases. Mutual influence of each phase on the drawing process is not excluded, and separation of the matrix polymer at the interphase boundary is possible. This is especially likely when a significant difference between the limiting drawing ratios of the components exists. The film surface may remain smooth or become rough, with a certain topography of roughness. All this depends on conditions of drawing: temperature, degree and rate of drawing.
3. The modulus of the dispersed phase is much higher than that of PP. In this case the dispersed phase separates from the PP matrix. When the difference in modulus is very high, the particles do not deform and the dispersed polymer does not orient. Surface roughness certainly appears. New optical effects can be revealed in the film, depending on the conditions of drawing. The film appears to be 'gas filled', with good light reflectance and metallic lustre as a consequence.

The drawing temperature has a great influence on the process of drawing and thus on the film properties. The temperature coefficients of the moduli of different polymers are usually different, which is the reason why the modulus ratios of both blend constitutents may change and thus a blend of group (1) may be deformed according to the mechanism of group (2) or even of group (3). For example, PP films with dispersed polystyrene deformed at 30 °C undergo matrix–particle separation and the density decrease is associated with increasing turbidity. The same films drawn at 120–140 °C stay almost transparent, without phase delamination, and their density changes linearly with the polymer composition ratio.

The morphology of the film changes drastically during drawing. The form and size of particles are very different in films from the same composition drawn under different conditions. Film samples are usually produced by extrusion when disperse particles become anisometric. It is very important to choose the right direction of the subsequent drawing for such films: along the direction of extrusion or in the perpendicular direction.

The following section illustrates the ideas discussed above.

5.3.1 Film drawing of PP blends with elastic polymer

The tensile strength of iPP film decreases after the addition of small amounts of any rubber or even of atactic PP (aPP) with molecular mass of limit $(20–30) \times 10^3$ g/mol (Figure 5.3) [10]. Density and tensile strength decrease with increasing aPP content. Uniaxial film drawing of an iPP and aPP mixture up to $\lambda = 12$ provides samples with higher strength but the negative effect of the aPP is still quite noticeable (Figure 5.4). It is also well known that this drawing ratio is not limiting for PP films. It is important

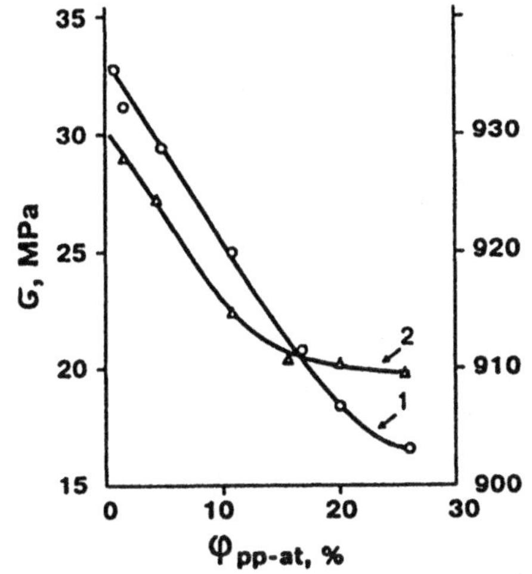

Figure 5.3 Dependence of (1) tensile strength σ and (2) density ρ on (aPP) content for nonoriented PP films.

to note that PP films with a certain aPP content have higher deformability. This permits the drawing of blend films with a drawing ratio greater than $\lambda = 12$ and thus the achievement of a higher degree of orientation, improved modulus, strength, and so on. In Figure 5.4, σ and Δn of PP films with different aPP contents are shown. Each film was drawn up to λ_{max}, which is the maximum possible elongation under the chosen drawing conditions. A sharp maximum was observed at an aPP content of 8–12%. This maximum is the result of a higher drawing ratio which could be reached in the presence of aPP. This aPP amount obviously enhances the deformability and thus the orientation of the macromolecules. It is the result of an improved ability of macromolecules to be oriented in contact with thin layers of soft and deformable aPP. The effect of aPP content may be treated analogously with the effect of temperature: the macromolecules of iPP align more easily and the resulting oriented film has fewer defects and internal stresses frozen in. The soft aPP, due to its incompatibility with iPP in the crystalline state, is probably concentrated on the boundaries of crystallites. It decreases the intermolecular interaction and, at the same time, increases the mobility of the macromolecules, which helps the fibrils to align in the direction of drawing. Increasing aPP content leads to a decrease of the 'load-bearing' of iPP and even destroys its phase continuity. All of this decreases σ and Δn, as well as the elongation at break.

5.3.2 Drawing of PP blends with components of similar moduli

The addition of polymers with moduli close to that of PP such as high density polyethylene (HDPE) and low density polyethylene (LDPE) reveals

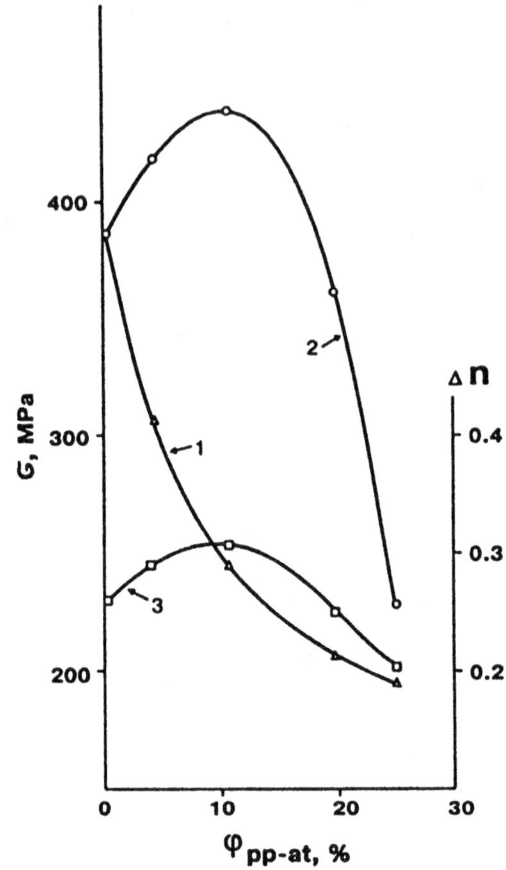

Figure 5.4 Dependence of (1, 2) tensile strength σ and (3) light birefringence Δn on (aPP) content for oriented PP films. Drawing ratio $\lambda = 12$ (1) and $\lambda = \lambda_{max}{}^{(2,3)}$.

a more complicated influence on the film properties of iPP [11–13]. So, addition of HDPE (75–78% crystallinity) to PP (crystallinity grade 56–58%) has a favourable influence on the mechanical properties of the film only at concentrations less than 15–20%. Figure 5.5 shows the relationship between σ and HDPE content for films prepared by pressing of mechanically mixed polymer powders. The σ value was calculated by considering the initial cross-section of the sample. It can be seen that the maximum on the curve in Figure 5.5 is not very significant. At the same time, the E modulus, determined at 1% elongation, as a function of the iPP–HDPE; composition, shows a rather sharp maximum. According to Figure 5.6 this maximum is located at about 20% HDPE.

It was shown [11] that the dispersed PE particles in the PP matrix reduced the PP spherulite size, so that the PE particles may be considered as nucleation agents. Increase of the PP matrix modulus is a consequence

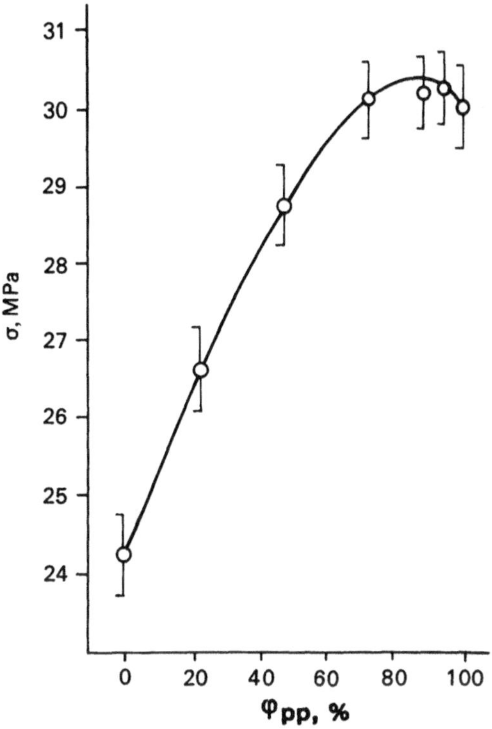

Figure 5.5 Tensile strength of HDPE–PP blend films versus PP content [11].

of the change in the PP crystalline structure due to the presence of highly crystalline PE particles. The positive effect of the addition of small amount of PE became significant only after orientational drawing of blend films [14]. It is followed by a sharp increase of the degree of PP orientation and also film stability to irradiation. Increase of thermostability and oxidation resistance is characteristic for drawn PP films with small amounts of PE added. The method of film preparation from blends affects the film properties. According to [12], PE matrix can be 'reinforced' by addition of PP only into the direction of the initial film extrusion. This effect can be explained by the 'armouring' effect of the partially oriented thin, fibrous, discontinuous PP phase.

Figures 5.7 and 5.8 show the relationship between macromolecular orientation and degree of crystallinity in blend films drawn up to $\lambda = 6$–7. Adding small amounts of PE to PP leads to a significant decrease of the PE crystallinity since the degree of orientation of the PE macromolecules increases. This may be as a result of PE macromolecules 'stretched' in the amorphous regions. It is confirmed by lowering of molecular mobility in the amorphous regions. Since PP phase orientation and crystallinity do not change markedly, it can be concluded that upgrading of the mechanical

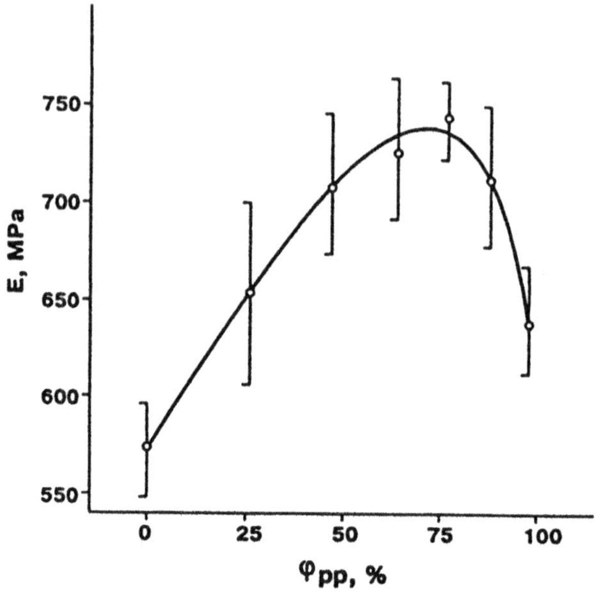

Figure 5.6 Elastic modulus at 1% elongation of HDPE–PP blend films versus PP content [11].

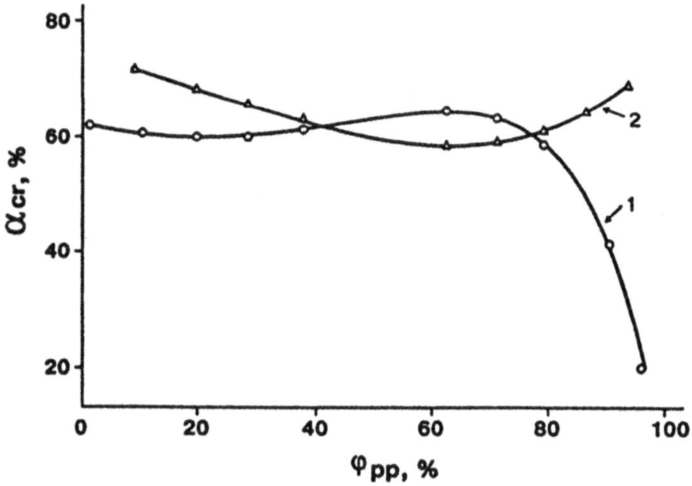

Figure 5.7 Dependence of degree of crystallinity on PP content in HDPE–PP blend films. 1, α_{cr} in HDPE; 2, α_{cr} in PP. Drawing ratio $\lambda = 6$–7. (According to [14]).

properties (especially modulus) is mainly due to the increment of the orientation of the macromolecules in the amorphous PE phase when a small amount of PE is added to PP. The addition of PP to PE does not cause significant changes in the morphology or in the mobility of 'structural elements' except an increase in crystallinity. All these effects are attributed to

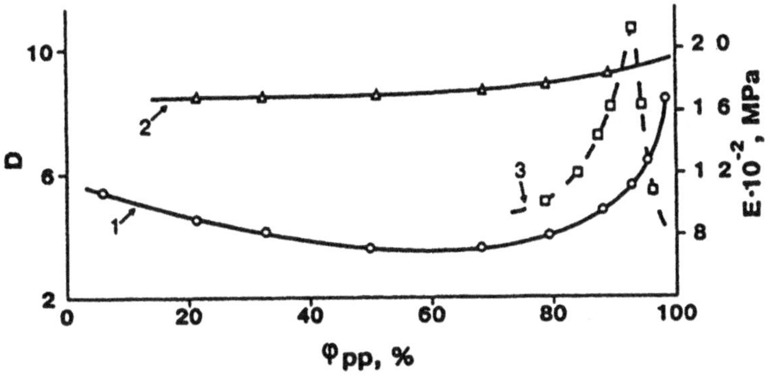

Figure 5.8 Dependence of (1, 2) dichroism factor *D* (1, 2) and (3) modulus *E* on PP content in HDPE–PP blend films. 1, *D* for HDPE; 2, *D* for PP [14].

strong interphase PP–PE interaction. PE segments penetrate into the interfibrillar regions of the PP phase and preserve their pronounced orientation during the process of PP phase deformation [14].

One must also take into consideration the different deformability of the polymer components of the blend during drawing. So, if λ_{max} for PP and HDPE are essentially the same, their blend is also characterized by a similar value of λ_{max} (Figure 5.9). Most interesting are the blends in which λ_{max} of PP and LDPE are different. As shown in Figure 5.9 λ_{max} of the blend films in this case differs from that calculated by the additive rule. The inflection point on the λ_{max}–composition curve shows the composition interval in which phase reversion of heterophase mixture occurs.

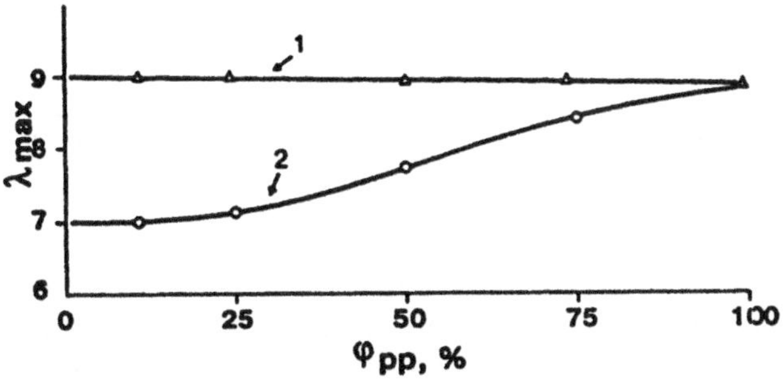

Figure 5.9 Dependence of λ_{max} on PP content in films composed of (1) HDPE and (2) LDPE.

The most important peculiarity of the orientational drawing of PP–PE blend compared to neat PP or HDPE drawing is the formation of roughness of a certain texture on the film surface. PP and HDPE films have insignificant roughness after the extrusion process and the smooth surface is preserved during the following drawing. In the process of blend drawing a blend with PP:PE ratio 50:50, as an example (Figure 5.10), the degree of roughness increases and its shape becomes irregular. Overhangs and indentations on the surface are a result of nonhomogeneous morphology and its nonhomogeneous deformation. An important aspect is that even in the rough film both polymer phases may stay continuous. The average surface roughness (R_a) increases with increasing drawing temperature. The most sharp increase of R_a occurs at $T > (T_m)$ of HDPE. At such conditions the PE melt forms 'insulating layers' between the crystalline entities of PP, which hinder the deformability of the latter. Despite this, no deterioration in tensile strength was observed. This is true for compositions with continuous PP phase. As can be taken from the above data, the PP phase in the blend is continuous up to rather low PP content. The relation between the blend composition and limiting temperature of drawing (T_{max}) is interesting. Like T_{max}, the highest possible temperature was taken at which the film could be drawn without tearing at the clamps.

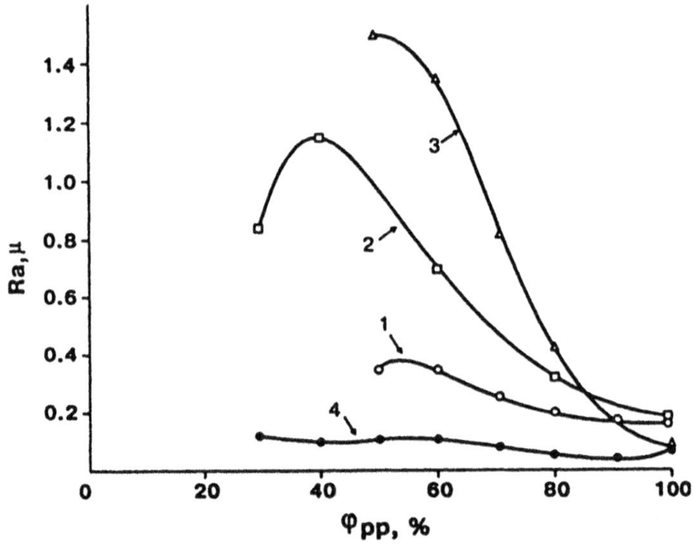

Figure 5.10 The relationship between surface roughness (R_a) and composition of PP–HDPE blend films. Drawing direction: (1, 3) transverse to the extrusion direction of the original film and (4) along the exrusion direction. Temperature of drawing: 1, 70 °C; 2, 130 °C; 3, 150 °C; 4, 80–130 °C.

The relationship between T_{max} and blend composition is shown in Figure 5.11. It can be seen from the data shown that the deformability of the blend film is determined by the polymer forming the continuous phase. The deformability of blend films at higher temperature is controlled by the PP phase with higher T_{max} value, which is continuous up to 15–20% of PP content. It is seen then, that PP incorporation to the rather low melting PE

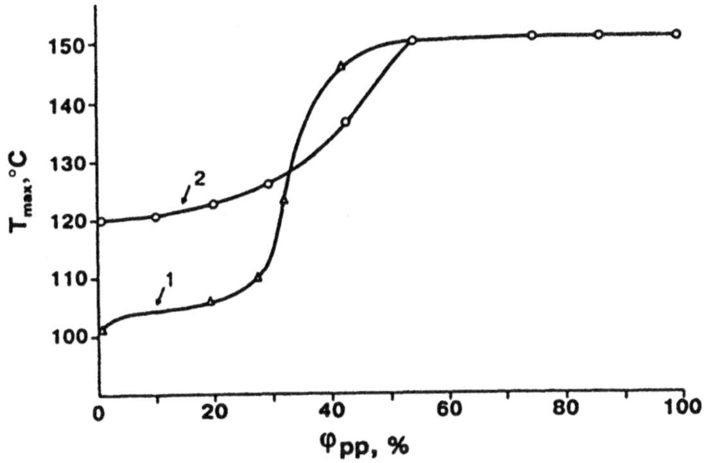

Figure 5.11 The relationship between limiting drawing temperature and PP content in blend films of (1) LDPE and (2) HDPE. Drawing rate: 1000%/min.

increases the heat stability, deformability and temperature interval of drawing of the films.

5.3.3 Film drawing of PP mixture with hard glassy polymers

PP blends with thermoplastics like polystyrene, polyamide and others are like mixtures of incompatible polymers with relatively sharp interphase boundary and weak interphase interaction [9]. Unoriented as well as oriented films from blends of PP with the above polymers have lower mechanical properties compared with those expected according to the additive rule. At the same time, such blend films impart some useful new properties as a result of drawing. It is important to note that the initial blend film structure has a strong influence on the film properties after drawing. So, the structure of extruded films of PP blend with up to 30% of PP in a poly(ethyleneterephthalate) (PET) matrix is actually a dispersion of PP spherical particles of 5–15 μm in the PET matrix [15]. The dispersed PP particles separate from the PET matrix at small deformations ($\lambda > 1.2$) and voids form during the process of drawing. Such film then become porous with closed cell character. At a drawing temperature which is optimal for PET orientation PP particles undergo elastic deformation only when the matrix becomes thinner after stretching. As a result the PP particles move towards the surface, and the film becomes rough. Places of overhang on the surface are PP particles covered by a thin layer of matrix PET. Along with the roughness, such films gain a silvery, dull colour because of multiple light reflection on the surfaces of particles and voids in the matrix. Extruded films of blends with PP matrix and PET disperse phase (10–30%) are characterized by a specific morphology of thin fibres and bands oriented in the extrusion direction. The length of the banded particles varies between 10 and 100 μm and their thickness is between 1 and 5 μm. The ultimate properties of oriented films with such a morphology depend strongly on the drawing direction: along or transverse to the extrusion direction of the initial film.

The influence of dispersed particles of elongated form on the properties of oriented films is more obvious in PP–polystyrene (PS) blend films [16]. For unoriented extruded blend films with 10% PS content the structure consists of ellipsoidal PS particles (short axis 0.8 μm), whereas in films with 50% PS content PS bands can be observed. Thin fibres of PP forms in PP–PS blend at 10% PP content. The fibre thickness is about 0.4 μm, the length reaching tens of micrometres. Nevertheless, macromolecules in such a highly asymmetric particle are not oriented substantially into the long axis direction of the particles. The deformability of nonoriented films with 80% PP content is very low (Figure 5.12). It can be seen that the elongation at break (ε_b) of such films is on the level characteristic for brittle PS. Rather unexpected sharp increase in the elongation at break of the blend films occurs after their drawing in the direction of the anisotropic particles' orientation (Figure 5.12). ε_b may reach 3–4 times the related value of oriented PS. This appears to contradict traditional experience, that ε_b decreases with increase of λ. The

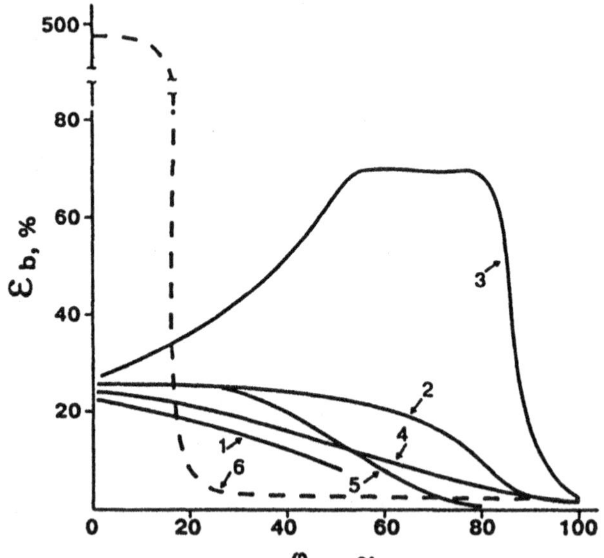

Figure 5.12 Elongation at break (ε_b) of films drawn at (1) 90 °C, (2, 4) 110 °C and (3, 5) 140 °C as a function of polymer ratio in PP–PS blend. Designations: 1, 2, 3, along the extrusion direction; 4, 5, transverse to the extrusion direction; 6, unoriented films. Drawing ratio, $\lambda = 5.5$.

presence of incompatible PS particles just provides additional structure defects and ε_b should decrease even more. This is true for nonoriented films. Blend films drawn at 110 °C have oriented PP matrix but they also have a number of PS particles as defects in the structure. At higher temperatures (140 °C) PS is in the viscoelastic state and consequently its Haw effect decreases. In a film drawn at 140 °C, the overall interphase are between PP and PS increases. The significant increase of the interphase layer volume as result of orientational drawing with low intermolecular interaction leads to high deformation ability of oriented films. The sharp decrease of ε_b at PS content greater than 80% is because of an interruption of the originally continuous PP phase. Film failure occurs as a result of PS brittle matrix deformation.

Drawing conditions influence the blend film properties, as shown in Figure 5.13. Films with average polymer ratio are oriented to a very low level compared to predicted one on the basis of light birefringence proportionality. The highest deviation from predicted values appears as a result of drawing at temperatures unfavourable for PP orientation (100–110 °C). PS particles, being hard at this temperature, serve as an obstacle for the orientation of PP macromolecules. At higher temperatures (such as 140 °C) PS particles are softened (*T* is higher than the glass transition temperature of PS) and do not prevent the orientation of PP macromolecules. In Figure 5.14 the Hermans factor of macromolecules orientation for the crystalline part of PP is presented as a function of PS content and drawing direction. From this data it follows that macromolecular orientation in the crystalline PP part occurs in the same way as in the unmodified PP film in a wide interval of the blend composition. However, the location of the oriented PP

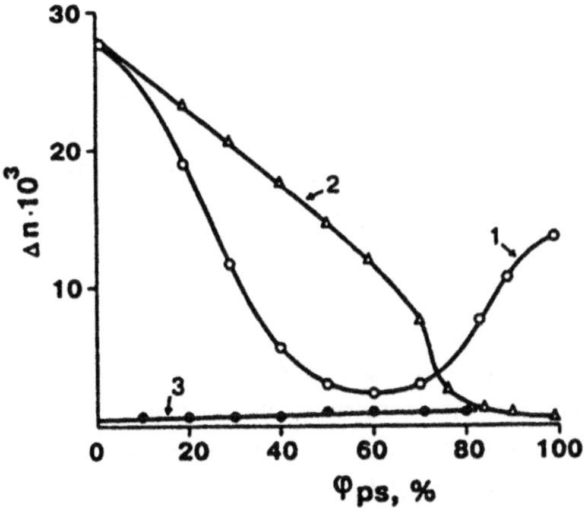

Figure 5.13 Dependence of light birefrigience (Δn) on composition of blend in PP–PS films. Films were drawn in the extrusion direction up to drawing ratio $\lambda=5.5$. Designations: drawing temperature: (1) 100 °C and (2) 140 °C; 3, unoriented films.

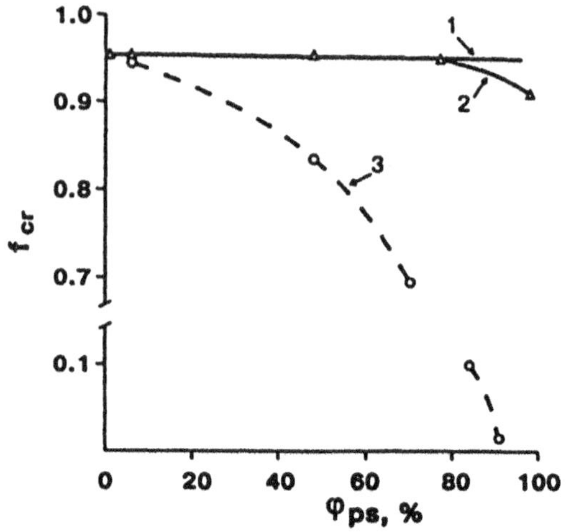

Figure 5.14 Relation between Hermans orientation factor (f_{cr}) in the crystalline PP part on the PS content in PP–PS blends. Drawing ratio $\lambda = 5.5$; drawing temperature (1, 2) 110 °C and (3) 140 °C. 1, drawing along the extrusion direction; 2, 3, drawing transverse to the extrusion direction.

fibres in PS matrix depends on the drawing direction of the original film: along or transverse to the extrusion direction (compare curves 1 and 3 in Figure 5.14). Drawing in the extrusion direction gives high PP orientation in a broad interval of blend composition, while after transverse drawing the f-factor decreases with decreasing PP content. At the same time, films of balanced compositions (composition ratio = 1:1) not only have different macromolecular orientation in crystallites but also different orientation in the crystallites themselves. So, in the film of PS:PP (1:1) blend drawn perpendicular to the extrusion direction only 30% of PP crystallites are oriented in the drawing direction. In the drawn films of PP:PS blend 1:9 ratio all PP crystallites are nonoriented ($f \to 0$). This change in f is determined by the blend morphology depending on the blend composition and the phase viscosity ratio of the components. Blend films with 10% PP content drawn in the extrusion direction at 140 °C contain ultrathin, long PP fibres. Hence PP macromolecules become oriented despite the softened state of the PS matrix, providing a high f-value of PP exists. Ultrathin PP fibres, essentially, are not connected with each other and rather high strength must be applied through the PS matrix to reorient them by drawing in the transverse direction. It is impossible to reach such stress levels at 140 °C because the PS matrix is very soft and the fibres remain essentially nonoriented (curve 3 in Figure 5.14).

The presence of fibre-like PP in PS matrix gives a significant change of film properties after orientation. It can be seen in Figure 5.15 that λ_{max} of the blend films with 10% PP is much higher than the value calculated by means

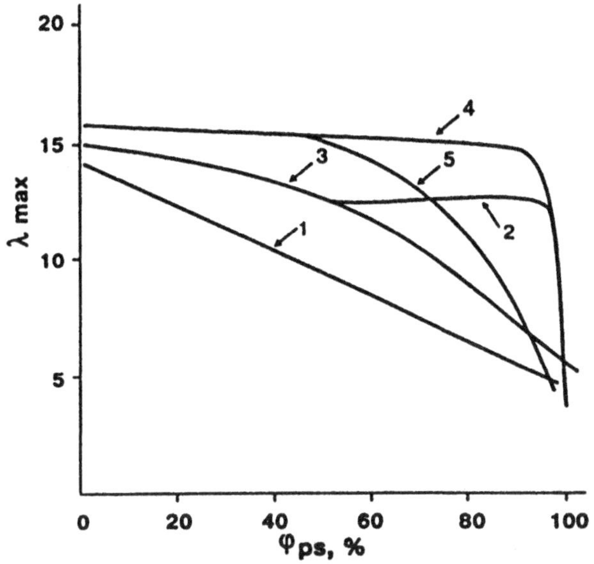

Figure 5.15 The relationship between limiting drawing ratio (λ_{max}) of oriented PP–PS blend films and PS content. Drawing temperature (1) 110 °C; (2) 130 °C; (4, 5) 150 °C. 1, 2, 4, along the extrusion direction; 1, 3, 5, transverse direction.

of the additivity principle. This is mainly because the fibrous PP phase maintains continuity (matrix partially) up to 10% PP content. Otherwise PP fibres form a volume skeleton which accepts applied stresses and prevents PS matrix failure at high drawing temperatures. Deformability of the blend films at lower drawing temperatures is also higher than the additive value but not as high as for films drawn at higher temperatures (Figure 5.15).

Optical effects as a result of drawing, such as a silvery and dull film surface, become more pronounced when drawing occures at rather low temperatures (90–100 °C), which are lower than T_g of PS. In this drawing condition only Hooke's deformation of PS particles occurs and consequently phase separation takes place only at the matrix–particle boundary. The density of such 'silvery' films is much less than estimated based on the additivity principle (Figure 5.16). The density decreases to a very low level if the film is drawn in the direction transverse to that of the extrusion. In this case phase separation occurs along the longer axis of anisometric PS particles and the overall volume of voids formed becomes higher.

Heterophase morphology with highly elongated anisotropic particles (fibres and bands) is rather similar to a laminated structure, which makes possible easy film separation on two or three stripes. Only very small stress is necessary to provide the delamination of PP–PS polyblend films (Figure 5.16). As an example, PP–PS blend films (PP:PS = 2:8) with thickness 120 μm drawn at 110 °C can be easily delaminated into two ~60 μm stripes.

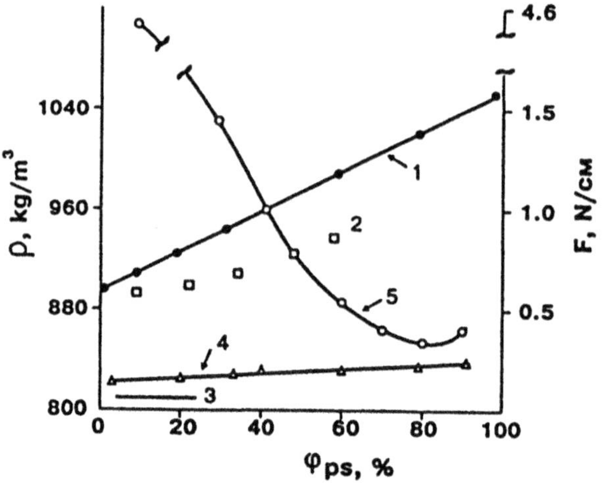

Figure 5.16 Dependence of (1, 3) density ρ and (4, 5) stress at which delamination on phase boundary occurs F on PS content in films of PP–PS blends. Drawing temperature: (1) 110–140 °C and (2, 3) 90 °C. Drawing ratio $\lambda = 5.5$; drawing direction: (1, 2) along and (3) perpendicular to the extrusion direction. 4, Unoriented films.

It is quite natural that the surface of the original film consisted mainly of PS because the latter predominates in the composition. Rather peculiar is that after delamination, the newly formed surface consist of both polymers in the same proportion as in the volume.

5.3.4 Film drawing of PP blends with polymers of alternating moduli

A polyolefin matrix may contain a second dispersed polymer, the modulus of which may be changed. This can be done by adding a selective plasticizer into the dispersed polymer, i.e. a plasticizer that is incompatible with the matrix and soluble only in the dispersed polymer. PP may be taken as a matrix polymer as well as HDPE because they do not swell in polar liquids, particularly at temperatures below the melting temperature. With such a system, the plasticizer can be introduced in the disperse phase and then gradually change the properties of the latter. This gives the opportunity to demonstrate changes in the blend properties and in the matrix due to orientation when only the feature of the dispersed polymer varies.

The relationship between properties of polyolefin–PS blend at a given amount of di-2-ethylhexyladipinate as plasticizer is shown in Figure 5.17 [17]. PP or HDPE was taken as a matrix: the type of polymer did not alter the general regularities with respect to the properties. The plasticizer had high selectivity towards PS: the degree of swelling of the polyolefin matrix is less than 3–5%, even at $T > T_m$.

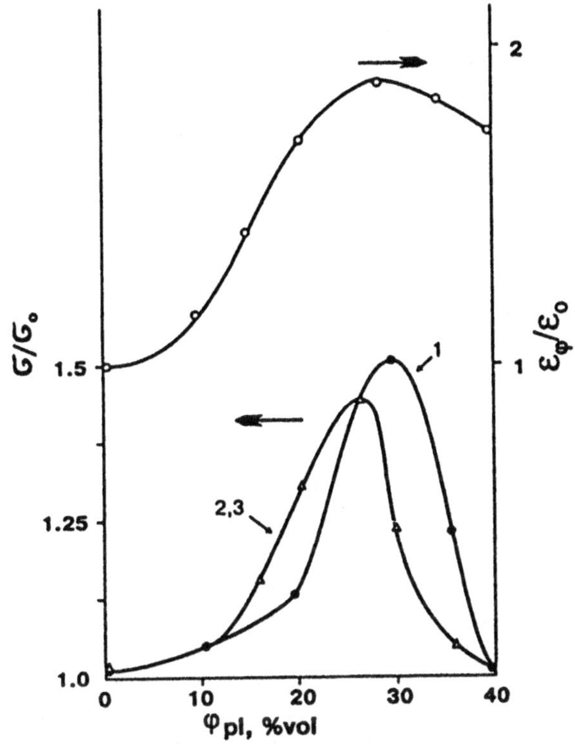

Figure 5.17 The relationship between relative value of tensile stress (σ/σ_0), elongation at break ($\varepsilon/\varepsilon_0$) and selective plasticizer content in the PS disperse phase for (1) PP, (2) PE and (3) PA films. PS content in film: 10 wt%. Drawing ratio, $\lambda = 6$–7. σ_0 and ε_0 are the stress and elongation at break without plasticizer, respectively.

It can be seen from Figure 5.17 that tensile strength and elongation at break values go through a maximum for the oriented films. The maximum is actually negligible for unoriented films. The macromolecular orientation degree (Δn) also increases with increasing plasticizer content. This effect can be explained in the same manner as in the case of aPP incorporation into iPP film. Glass transition temperature, modulus and viscosity of dispersed PS decrease after plasticizing. Such 'soft' particles elongate during orientational drawing into long fibres and bands: these pliable interlayers lead to an increased degree of orientation of the matrix polymer (PP or HDPE). The best effect occurs at 25–30 vol% plasticizer content in PS. Further decrease of PS viscosity leads to significant weakening of the blend film due to deterioration in the tensile strength. In the explanation of the above mentioned effect of elastic PS particles that are able to 'reinforce' PE matrix, we should also remember the interphase layer on the boundary between incompatible polymers. The thickness of this layer is higher, the more similar are the polymers in their chemical nature. The thickness of the interphase layer varies most probably between 1 and 10 nm. In the case of a crystalline polymer its amorphous part participates in the interphase layer formation. Interaction between segments of different polymers in this layer of 'segmental solubility' [9] is lower than in the volumes of each phase, occupied by like segments (macromolecules). From this we may conclude that the higher is the deformability of the dispersed phase particles, the greater is the internal surface which separates the two polymers and the higher is the general volume of blend in the interphase layer. This means that there are two reasons why the PE matrix deformability increases:

1. Mechanical (the presence of elastic, thin interlayers);
2. Greater amount of blend with reduced intermolecular interaction in the enlarged interphase layer.

Both reasons helps the relaxation of overstress that arise during drawing of the hard matrix.

The relative values of properties are shown in Figure 5.17. They are compared to properties of blends without a plasticizer. Such an approach makes it possible to demonstrate that the data for PP–PS and HDPE–PS as well as for HDPE–PA-6 blends on the same curve. This coincidence of the experimental data underlines the general character of this effect.

We can then conclude from the experimental results described above that the blend constituents influence the orientational drawing of PP significantly. Special factors are: method of original film preparation, phase morphology of the heterogeneous blend film, conditions of the drawing process and direction of drawing with respect to the initial film extrusion direction. The presence of another polymer strongly affects the efficiency of the macromolecular orientation of the PP matrix. After orientational drawing PP-based blend films show a wide variety of new properties which make them useful for many industrial, agricultural and domestic applications.

5.4 APPENDIX: SYMBOLS AND ABBREVIATIONS

aPP	Atactic PP
f	Macromolecular orientation factor (Hermans factor)
iPP	Isotactic PP
PA-6	Polyamide
PET	Poly(ethyleneterephthalate)
PP	Polypropylene
PPhTA	Poly-(p-phenyleneterephthalamide)
PS	Polystyrene
R_a	Roughness of the film surface
T_m	Melting temperature
T_{max}	The highest temperature at which a film does not tear at clamps
Δn	Light birefringence
ε_b	Elongation at break

λ Drawing ratio (l/l_0)
λ_{max} Limiting drawing ratio
σ Tensile strength at break

5.5 REFERENCES

1. Erina, N. A., Knunianz, M. I. and Dorfman, I. Ya. *et al.* (1985) *Dokladi Akademii Nauk SSSR*, **280**(4), 913–16.
2. Unwin, A. P., Bower, D. J. and Ward, I. M. (1985) *Polymer*, **26**, 1605–10.
3. Liu Tuo-min, Guska, T. D. and Harrison, J. R. (1986) *Polymer*, **27**, 247–49.
4. Vlassov, S. V. and Sagalaev, G. V. (1969) *Mechanika polimerov* (*Polymer Mechanics*), **4**, 735–8.
5. Savatari, Ch. and Matsuo, M. (1986) *Macromolecules*, **19**, 2633–56.
6. Capaccio, G., Gibson, A. G. and Ward, I. M. (1979) in A. Cifferi and *Ultra-High Modulus Polymers*, (eds I. M. Ward), Applied Science Publishers, London.
7. Myasnikov, G. D. and Sergunko, M. A. (1971) *Plastmassi* (*Plastics*), **1**, 30–31.
8. Nordmeier S., Menges, G. (1986) *Advances in Polymer Technology*, **6**, 59–64.
9. Kuleznev, V. N. (1980) *Polymer Mixtures: Structure and Properties*, Chemistry Publishing House, Moscow, pp. 190–4.
10. Sagalaev, G. V., Vlassov, S. V., Lekae, I. A. and Krochina, V. V. (1976) *Plastmassi* (*Plastics*), 11, 44–7.
11. Lovinger, A. and Williams, M. (1980) *Journal of Applied Polymer Science*, **25**, 1703–13.
12. Sherman, E. S. (1984) *Journal of Materials Science*, **12**, 4014–20.
13. Atanassov, A. and Nedkov, E. (1984) *Bulgarian Journal of Physics*, **11**(2), 203–17.
14. Karpov, S. G., Popov, A. A. and Zaikov, G. E. (1981) *Vysokomolekulyamie Soedineniya* (*High Molecular Compounds*), **A33**(12), 2585–93.
15. Vlassov, S. V., Berezhnaya, G. V. and Kuleznev, V. N. (1981) *Plastmassi* (*Plastics*), 4, 29–30.
16. Heinemann, P., Vlassov, S. V., Miroshnikov, Y. P. and Kuleznev, V. N. (1986) *Kolloidnii Jurnal* (*Journal of Colloid Science*), **48**, 974–9.
17. Kuleznev, V. N. and Vlassov, S. V. (1989) *Makromolekulare Chemie, Macromolecular Symposia*, **28**, 85–105.

6

Thermoplastic elastomers by blending and dynamic vulcanization

A. Y. Coran and R. P. Patel

6.1 INTRODUCTION

Thermoplastic elastomers (TPEs) are a class of polymeric materials that have many of the performance-related properties of conventional thermoset elastomers but are melt processable as thermoplastics. Over the years, TPEs have evolved to become high performance TPE materials. Today they are available with a wide range of end use properties.

The earlier classes of TPEs were based on block copolymer technology. The development of TPEs by that technology required expensive reactors and was limited to instances where hard and soft blocks could be combined in a single molecule.

In recent years, another class of elastomeric products based on rubber–plastic blends has become technologically interesting for use as thermoplastic elastomers [1–3]. This is because one can select soft (rubbery) and hard (resinous) components of the blend, from a wide variety of polymeric products. These blend-based products also possess many of the properties of thermoset rubbers, yet they are processable as themoplastics [4]. They do not need to be vulcanized during fabrication into end-use parts. Conventional thermoset fabrication technology is compared with the thermoplastic elastomer fabrication technology in Figure 6.1.

The TPE fabrication process is simple, economical and efficient. It also offers recyclability of the scrap/waste TPE materials. Thus, TPEs offer a substantial economic advantage with respect to the fabrication of

Polypropylene: Structure, blends and composites. Edited by J. Karger-Kocsis.
Published in 1995 by Chapman & Hall, London. ISBN 0 412 58430 1

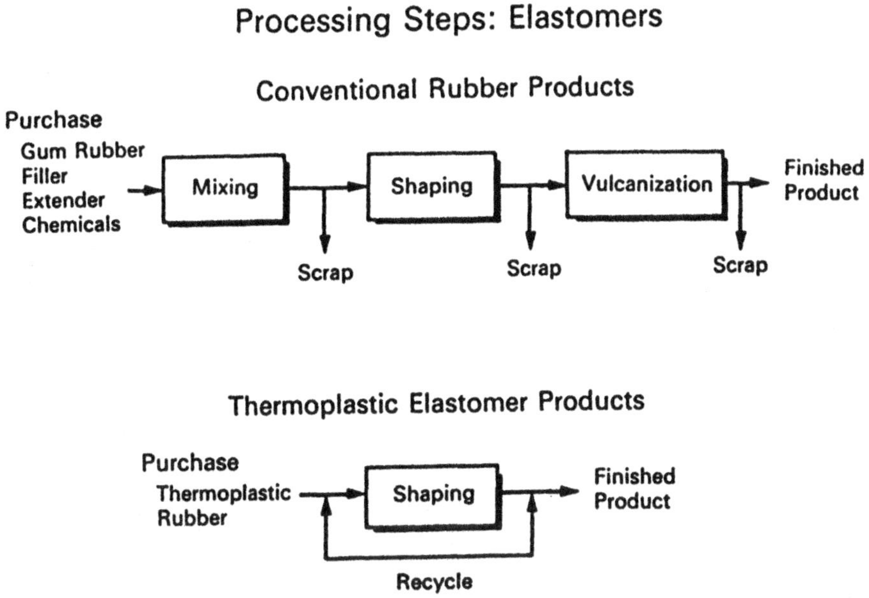

Figure 6.1 Comparison of thermoset and thermoplastic processing steps.

finished-part production. The lower cost of thermoplastic processing has been the motivational spirit for the development of thermoplastic elastomers. This has led to a higher growth rate for the use of thermoplastic elastomers as compared to thermoset elastomers, as illustrated by Figure 6.2 and 6.3.

Although the number of possible elastomeric blend compositions is quite large, relatively few of them have been of technological importance. This is largely because of the fact that most polymers have been incompatible with one another, at least in the thermodynamic sense [5]. If the polymers of a blend are thermodynamically compatible, then their blend could exist as a single phase, with mixing having been accomplished on the molecular-scale level. In such a case, the properties of the resulting blend would tend to be the average of the properties of the 'pure' phases. For example, blending a rubber with a thermodynamically compatible plastic would give a composition with a glass transition temperature between those of the rubber and the plastic [6, 7]. Frequently, such an average glass transition temperature is near room temperature. Thus, a rubber and plastic might be blended to give a highly damped, almost 'leathery' material. This, of course, would not be a desirable product for use as a thermoplastic elastomer.

On the other hand, since most of the rubbers and plastics are thermodynamically incompatible, their blends contain two phases, with two glass transition temperatures. Such blends frequently contain large particles of one of the polymers only loosely bonded (if at all) to the other (matrix) polymer [8]. The large, essentially unbonded particles generally act as

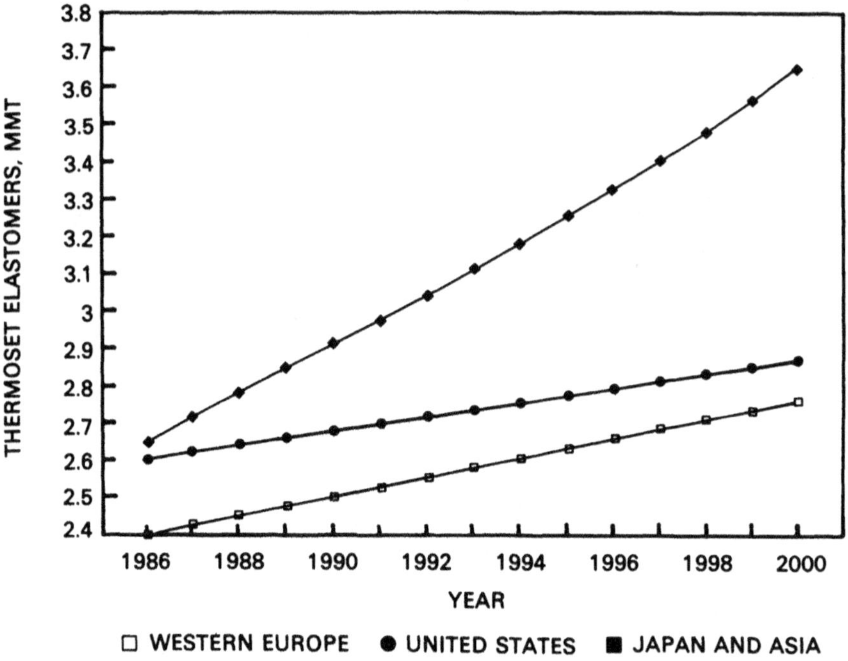

Figure 6.2 Projected growth in thermoset elastomer consumption.

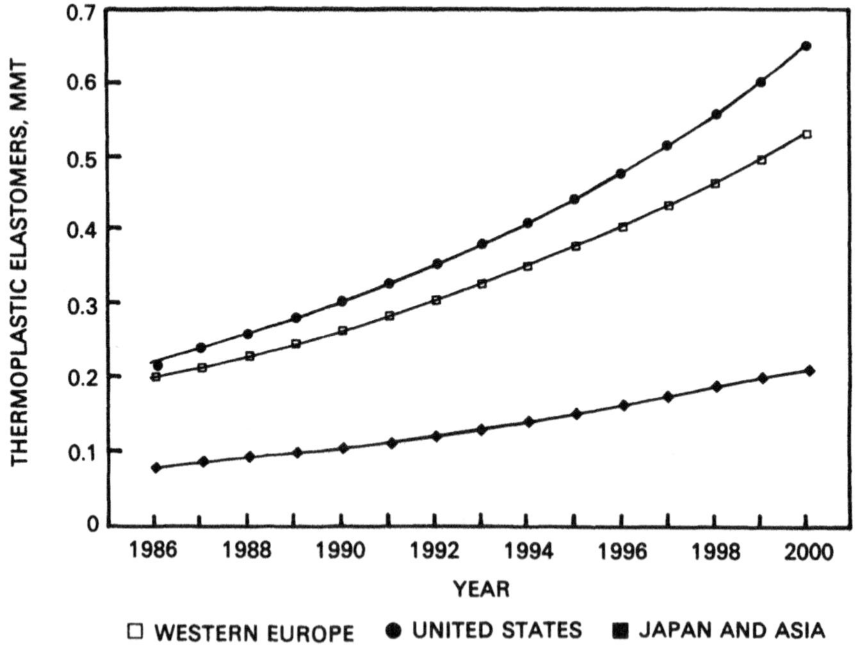

Figure 6.3 Projected growth in thermoplastic elastomer consumption.

stress-concentrating flaws. In some cases it is possible to obtain very small droplets of one polymer dispersed in the other during mixing; however, coarsening of the droplets can occur by coalescence, during some phase of processing a blend into a fabricated part [9]. Usually such compositions with large particles and poor phase adhesion have poor performance properties.

In a few instances, where, polymers approach mutual compatibility, it is possible to create technologically important materials. For many technological end-use applications, the ideal elastomeric rubber–plastic blend would comprise finely divided rubber particles dispersed in a relatively small amount of plastic. The rubber particles can be crosslinked to promote elasticity (the ability of the blend composition to retract forcibly from a large deformation). The favorable dispersed phase morphology should remain during the fabrication of the material into parts to insure sustained performance.

It should be pointed out that in the 'ideal' case proposed above, many of the desired properties could arise as a result of the polymers being almost, but not quite, thermodynamically compatible. The low glass transition temperature of the rubber phase (not 'averaged up' by the hard phase material) would be maintained because of the relative purity of the rubber phase; yet the high crystallinity or high glass transition temperature of the hard plastic phase could be retained for structural integrity. Examples of materials which approach the ideal case include compositions of crystalline polyolefin plastics containing domains of crosslinked polyolefin rubber (e.g. polypropylene (PP) with ethylene–propylene diene terpolymer (EPDM)).

At least in principle, a very large number of rubber–plastic blends are possible [10, 11]. This is because of the wide variety of commercially available rubbers and plastics that could be blended over a wide range of rubber–plastic proportions. However, in this chapter, we will be concerned with PP-based compositions which are soft or flexible and which have rubber-like elasticity in that they retract forcibly from large deformations. Such compositions usually contain larger proportions of rubber than do impact resistant, rubber-toughened plastics [12], which will not be considered here.

Excellent thermoplastic elastomeric compositions comprising vulcanized rubber particles in melt-processable plastic matrices can be prepared by dynamic vulcanization. Dynamic vulcanization is the process of vulcanizing rubber during its melt-mixing with molten plastic. This chapter describes the dynamic vulcanization process and the PP-based thermoplastic elastomeric compositions produced thereby. In addition, the scope of applicability of dynamic vulcanization is discussed. Also included is a discussion of end-use applications of some of the materials that can be prepared by this method.

6.2 HISTORY

Gessler [13] prepared 'semi-rigid' rubber–plastic compositions containing minor proportions of vulcanized rubber by mixing the rubber (chlorobutyl rubber), plastic (PP) and vulcanizing agent together and then, with continuing mixing, heating the mixture to partially vulcanize the rubber and to melt the PP. He called the process dynamic vulcanization, but it was different to the process as defined here because the polymers were not well blended in the molten state before the onset of vulcanization.

Fischer [14] used a dynamic vulcanization process to prepare compositions containing varying amounts of partially vulcanized EPDM rubber and PP. Large proportions of rubber were generally used and soft compositions were obtained. However, an organic peroxide was used to crosslink the rubber in the presence of PP, which was greatly damaged by the action of the peroxide.

More recently, it has been found that very strong, elastomeric compositions of EPDM and PP be prepared by dynamic vulcanization, provided that peroxide curatives are avoided [15, 16]. If enough plastic phase is present in the molten state, then the compositions are processable as thermoplastics. Plasticizers or extender oils can be used to expand the volume of the rubbery phase. In the molten state, the plasticizer can expand the volume of the plastic or 'hard' phase. If the hard phase material is a crystalline material such as PP, then, upon cooling, the crystallization of the hard phase material can force the plasticizer out of the hard phase into the rubbery phase. Thus, the plasticizer is a processing aid at melt temperatures but a softener at the lower temperatures of use. Although the dynamic vulcanization technique has been applied to many different plastic–rubber blends to prepare a wide range of thermoplastic elastomer compositions, only those compositions in which PP has been used will be discussed here.

6.3 THE PREPARATION OF THERMOPLASTIC ELASTOMERS FROM BLENDS BY DYNAMIC VULCANIZATION

Polymer blends, in general, have been prepared commercially by melt mixing, solution blending, and latex mixing [17]. Rubber–plastic blends of the type discussed here (those containing rather large amounts of rubber) have generally been prepared by melt-mixing techniques. Melt-mixing avoids problems of contamination, solvent or water removal, etc. In general, Banbury mixers, mixing extruders and twin-screw mixers are suitable for melt-mixing rubbers with plastics. However, for the purposes of this discussion, emphasis will be on laboratory melt-mixing techniques, which simulate what can be done in a factory.

The procedures given below are based on the use of either a small Brabender mixer or a Haake Rheomix. In each case the mixer was fitted with cam-type rotors, and optimum batch sizes were between 55 and 65 g.

The dynamic vulcanization process, which is applicable to many rubber–plastic combinations, can be described as follows: rubber, plastic and other ingredients (filler, plasticizer, lubricant, stabilizer etc.) are first melt-mixed. After sufficient molten-state mixing to form a well-mixed blend, vulcanizing agents (curatives, crosslinkers) are added. Vulcanization then occurs while mixing continues. The more rapid the rate of vulcanization, the more rapid the mixing must be to insure uniform crosslinking of the blend composition. It is convenient to follow the progress of vulcanization by monitoring mixing torque or mixing energy requirement during mixing process. After the mixing torque or mixing energy curve goes through a maximum, mixing can be continued somewhat longer to improve the fabricability of the blend. If desired, one can add some of the ingredients after the dynamic vulcanization is completed. After discharge from the mixer, the blend containing vulcanized rubber and the thermoplastic can be milled, chopped, extruded, pelletized, injection molded, or processed by any other desirable technique.

6.4 THE EFFECT OF DYNAMIC VULCANIZATION ON PROPERTIES OF THERMOPLASTIC–RUBBER BLENDS

As we stated earlier, dynamic vulcanization is the process of vulcanizing rubber during its intimate melt-mixing with a non-vulcanizable thermoplastic polymer. The rubber phase is vulcanized under shear to give a particulate vulcanized rubber phase morphology. The vulcanized rubber particles remain stable during melt processing of the TPE. The dynamic vulcanization of rubber–plastic blends produces compositions that have the following property improvements compared to similar compositions that are not vulcanized:

- reduced permanent set
- improved ultimate mechanical properties
- improved fatigue resistance
- greater resistance to attack by fluids, e.g. hot oils
- improved high-temperature utility
- greater stability of phase morphology in the melt during fabrication
- greater melt strength
- more reliable thermoplastic fabrication

In short, dynamic vulcanization can provide compositions that are more rubber-like in their end-use performance characteristics, and can be more rapidly processed to a final part than thermoset rubbers.

Compositions of greatly improved permanent set can be produced by even slight or partial vulcanization of the rubber. Such slightly vulcanized compositions can be produced by the partial vulcanization of the rubber before it is mixed with the plastic or by dynamic vulcanization of the plastic–rubber blend [14]. However, the other improvements can only be obtained (at least in the case of EPDM–polyolefin compositions) by dynamic vulcanization in which the rubber is technologically fully vulcanized.

6.5 PP-BASED THERMOPLASTIC ELASTOMERS

6.5.1 EPDM–PP-based thermoplastic elastomers

The dynamic vulcanization of blends of EPDM rubber with PP and with polyethylene (PE) have been described by Coran and Patel [16]. Physical properties, such as hardness, tension set values, and other parameters associated with unfilled compositions are given in Table 6.1. The general recipe for the blend compositions was as follows:

EPDM (Epsyn® 70A)	100
Polyolefin resin	X
Zinc oxide	5
Stearic acid	1
Sulfur	Y
Tetramethylthiuram disulfide (TMTD)	Y/2
2-Benzothiazolyl disulfide (MBTS)	Y/4

where X is the number of parts by weight of polyolefin resin, and Y is the amount of sulfur.

Not all of the compositions of Table 6.1 were prepared by dynamic vulcanization. In the first four compositions, the rubber was first press-cured and then ground, by tight roll-milling, to various particle sizes. The ground rubber particles were then mixed with molten PP.

Particle sizes were determined by optical microscopy. For the measurement of the crosslink density of the rubber, samples of the rubber alone were press-cured under conditions selected to simulate the conditions of dynamic vulcanization. Crosslink densities of the press-cured samples were then determined on the basis of solvent-swelling measurements by using the Flory–Rehner equation [18]. Mechanical properties and hardness measurements were made by the usual means.

The effect of the rubber particle size on the stress–strain properties of the EPDM/PP-based compositions is shown Figure 6.4. This is a composite stress–strain curve constructed from the data associated with compositions 1–5 of Table 6.1. Each × denotes a stress and strain at rupture or failure. From Figure 6.4, ultimate strength, ultimate elongation and energy to break (area under the curve from origin to an appropriate ×) are apparent. The average rubber particle size associated with each × is noted. The ultimate properties vary inversely with the diameter of crosslinked rubber particle.

Major effects of changes in crosslink density are given by Figure 6.5. Only a small amount of crosslink formation is required for a large improvement

Table 6.1 Properties of unfilled thermoplastic elastomer compositions

Composition number	Resin type[a] (phr[b])	Sulfur (phr)	Method of preparation[e]	Crosslink density 'moles' × 10^5/ml	Rubber particle size (μm) d_n	Rubber particle size (μm) d_w	Shore hardness, D	Young's modulus (MPa)	Stress at 100% strain (MPa)	Tensile strength (MPa)	Ultimate elongation (%)	Tension set (%)
1	PP/66.7	2.00	S	16.4	72	750	43	97	8.4	8.6	165	–
2	PP/66.7	2.00	S	16.4	39	290	41	102	8.4	9.8	215	22
3	PP/66.7	2.00	S	16.4	17	96	41	105	8.4	13.9	380	22
4	PP/66.7	2.00	S	16.4	5.4	30	42	103	8.4	19.1	480	20
5	PP/66.7	2.00	D	16.4	about 1–2		42	58	8.0	24.3	530	16
6	PP/66.7	1.00	D	12.3	–		40	60	7.2	18.3	490	17
7	PP/66.7	0.50	D	7.8	–		39	61	6.3	15.0	500	19
8	PP/66.7	0.25	D	5.4	–		40	56	6.7	15.8	510	19
9	PP/66.7	0.13	D	1.0	–		35	57	6.0	9.1	407	27
10	PP/66.7	0.00	–	0.0	–		22	72	4.8	4.9	190	66
11	PP/33.3	1.00	D	12.3	–		29	13	3.9	12.8	490	7
12	PP/42.9	2.00	D	16.4	–		34	22	5.6	17.9	470	9
13	PP/53.8	2.00	D	16.4	–		36	32	7.6	25.1	460	12
14	PP/81.8	2.00	D	16.4	–		43	82	8.5	24.6	550	19
15	PP/122	2.00	D	16.4	–		48	162	11.3	27.5	560	31
16	PP/233	5.00	D	14.5	–		59	435	13.6	28.8	580	46
17	None[c]/0	2.00	S	16.4	–		11	2.3	1.5	2.0	150	1
18	PP[d]/∞	0.00	–	–	–		71	854	19.2	28.5	530	–
19	PE/66.7	2.00	D	16.4	–		35	51	7.2	14.8	440	18
20	PE/66.7	0.00	–	0.0	–		21	46	4.1	3.5	240	24

[a] PP, polypropylene, Profax®6723; PE, polyethylene, Marlex®EHM 6006.
[b] phr; parts per 100 parts of rubber.
[c] Control composition purely of cured rubber.
[d] Control composition purely of PP.
[e] S, static (conventional) vulcanization; D, dynamic vulcanization.

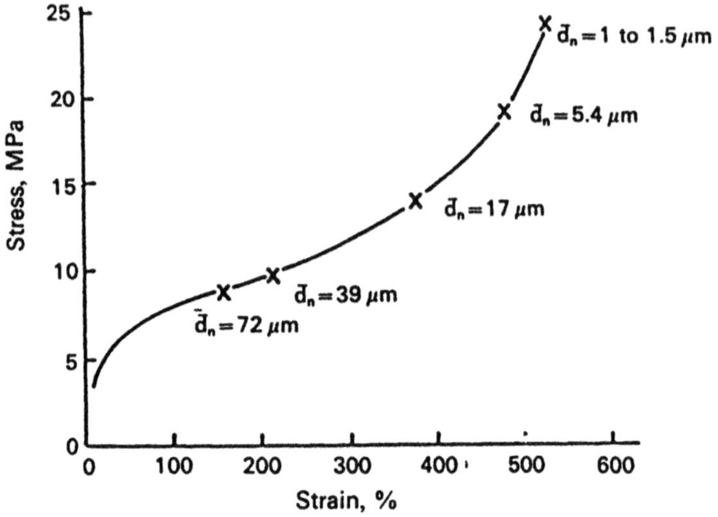

Figure 6.4 The effect of vulcanized rubber particle size on mechanical properties of EPDM–PP-based TPEs (× denotes failure).

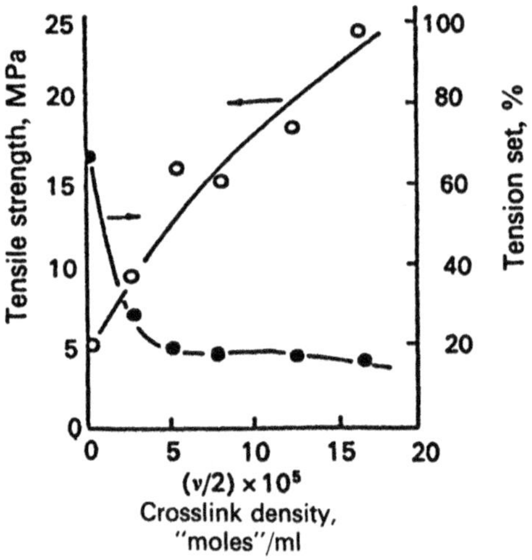

Figure 6.5 The effect of crosslink density on tensile strength and tension set of EPDM–PP-based compositions.

in tension set. Tensile strength improves continously as the crosslink density of the rubber phase increases, and the compositions remain capable of forming shaped articles as thermoplastics even at high rubber crosslink densities. Only small changes in the stiffness of the compositions occur with large changes in the extent of cure.

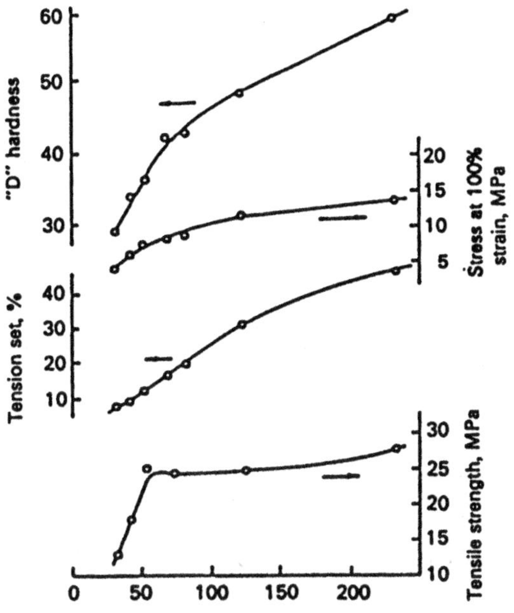

Figure 6.6 The effect of polypropylene content on EPDM–PP based thermoplastic elastomers.

As the amount of PP resin increases and the amount of rubber decreases, the compositions become less like the rubber and more like the plastic (Figure 6.6). Modulus, hardness, tension set and tensile strength increase. The shape of the plot of tensile strength versus the proportion of PP is interesting. Tensile strengths are low until at least 30 parts of polyolefin resin per 100 parts (phr) of rubber are used. Then, as the proportion of polyolefin resin is further increased, strength rapidly increases until about 50 phr of polyolefin. Further increases in the amount of resin can increase the tensile strength only slightly.

Shear modulus, as a function of temperature, is shown in Figure 6.7. The composition is the same as number 5 in Table 6.1. The dynamic shear moduli were determined by means of a torsion pendulum [19]. An effect of dynamic vulcanization is to prevent a complete loss of elasticity and strength at the melting point of PP. The vulcanized composition continues to exhibit sufficient strength for the torsion pendulum modulus measure-ments, even after the plastic phase melting point has been surpassed.

Another effect of dynamic vulcanization, indicated by the results of Figure 6.7, is a smaller decrease in modulus as the temperature is increased. The unique performance of the thermoplastic vulcanizates even above their melting points will indicate a variety of high temperature applications.

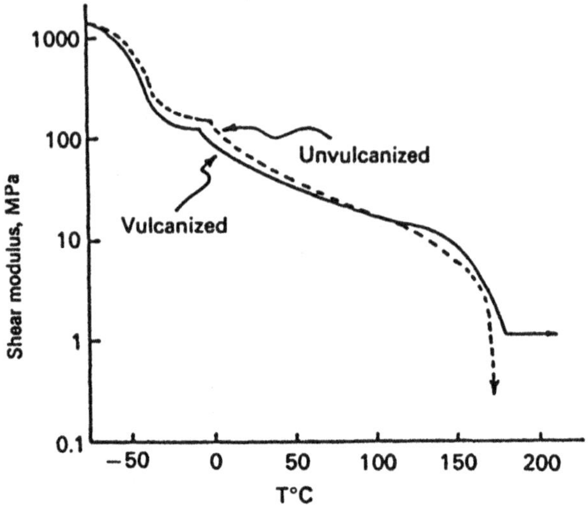

Figure 6.7 The effect of temperature on the stiffness of EPDM–PP blend compositions.

The data given for compositions 19 and 20 of Table 6.1 indicate that the tensile properties of the EPDM–PE blends are also greatly improved by dynamic vulcanization. However, on the basis of comparisons between compositions of PE with those of PP one concludes that the best compositions are prepared from PP.

A given rubber can be vulcanized by many different curing agents or their combinations. In addition to crosslink density, the type of curative also has a marked effect on the TPE end-use properties (Table 6.2). Abdou-Sabet and

Table 6.2 Effect of crosslinking agent on properties of EPDM–PP-based (60:40) TPEs

		Crosslinker		
Property	*None*	*Sulfur*	*Dimethylol phenolic type*	*Peroxide*
Tensile strength (MPa)	5.00	24.3	25.6	15.9
Stress at 100% strain (MPa)	4.80	8.0	9.7	8.1
Ultimate elongation (%)	190	530	350	450
Shore hardness, D	36	43	44	39
ASTM # 3 oil swell(70 h at 125 °C) (wt%)	Disintegrated	194	109	225
Compression set (22 h at 110 °C) (%)	91	43	24	32

Fath [20] have shown that when dimethylol phenolic resin is used as a curative, the EPDM–PP based TPEs give improved (lower) compression set values and oil swell as compared to other curatives. There were also improvements in processibility. These improvements were key to the commercialization of these materials.

The data of Table 6.3 demonstrate the effects of carbon black loading and oil extension. The compositions are variations of number 15 of Table 6.1. The effect of the filler is to strengthen the composition somewhat and to give some stiffening (with respect to hardness and stress at 100% strain, but not with respect to the 'zero strain' Young's modulus). Both oil extension and carbon black loading can give compositions with lower cost and excellent quality.

From data such as those in Table 6.3, and from measurements of hot oil resistance, fatigue life, etc., it has been concluded that oil-extended, filled compositions can be prepared, which are thermoplastic but which perform very much in the same way as conventional thermoset rubber vulcanizates.

One very important attribute of this technology is that it allows one to upgrade the performance of partially crosslinked TPE by advancing the degree of crosslinking at a later stage. For example one can first prepare a partially crosslinked TPE. Then, at a later stage, if desired, one can further crosslink the rubber and improve its properties [15]. This is shown in Table 6.4, which compares the tensile properties of the original partially crosslinked composition to the properties after further vulcanization.

One might expect that dynamic vulcanization would greatly affect the microphase morphology of rubber–plastic blends. This expectation is expressed by the schematic representation of the morphology of rubber–plastic blends, before and after dynamic vulcanization, given by Figure 6.8.

Compositions based on EPDM–PP blends were recently investigated by Abdou-Sabet and Patel [21] in order to gain an understanding of the effects of composition and degree of crosslinking on morphological changes and on the properties of the compositions. For uncured EPDM–PP blends, the morphology depended upon the EPDM:PP ratio (Figure 6.9). In the case of activated dimethylol phenolic resin cured blends, in the early stages of vulcanization, two co-continuous phases develop, even when well-defined uncrosslinked blends are used as starting material. As cure progresses, phase inversion and dispersed morphology were obtained (Figure 6.10). Here, at the end of the process, the rubber exists as vulcanized particles dispersed in a PP matrix.

EPDM–PP TPEs prepared with sulfur-based curatives also have dispersed phase morphology after dynamic vulcanization. However, sulfur

Table 6.3 Effect of carbon black and extender oil

Carbon black[a] *(phr)*	*Extender oil*[b] *(phr)*	*Tensile strength (MPa)*	*Stress at 100% strain (MPa)*	*Young's modulus, (MPa)*	*Ultimate elongation (%)*	*Shore hardness*	*Tension set (%)*
0.0	0.0	27.5	11.3	162	560	48D	31
80.0	0.0	31.0	14.3	120	410	51D	30
0.0	80.0	15.2	6.4	47	550	29D	19
80.0	80.0	23.0	7.2	23	530	33D	16
80.0	160.0	15.2	4.8	11.5	490	74A	13

[a] Carbon black N 327.
[b] Extender oil Sunpar™ 2280.

Table 6.4 Effect of advancing cure on the properties of EPDM–PP TPE

Composition	*Partially vulcanized TPE*	*Partially vulcanized TPE with additional vulcanization*[a]
Tensile strength (MPa)	6.77	14.70
Stress at 100% strain (MPa)	5.98	7.26
Ultimate elongation (%)	200	420
Gel content (%)	91	99

[a] Additional vulcanization was carried out under dynamic vulcanization conditions using a sulfur-based curative package.

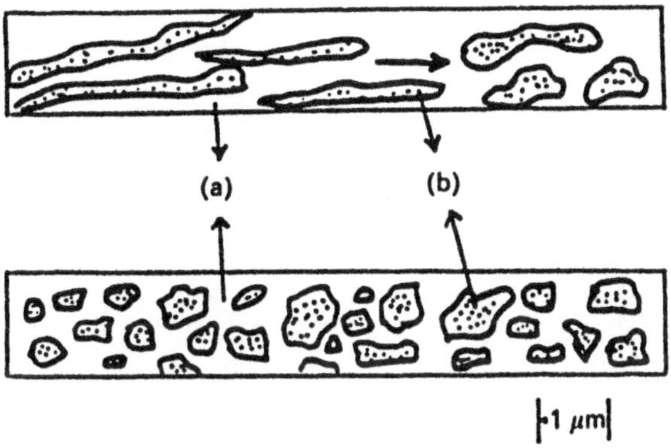

Figure 6.8 Schematic representation of crosslinked-particulate polymer-filled composites. (a) Continuous phase; (b) discontinuous phase.

links are capable of exchanging among themselves, a capability that can cause the rubber particles to agglomerate into larger, continuous domains (Figure 6.11). This, in turn, can lead to loss of fabricability for the TPEs.

6.5.2 Diene rubber-PP based thermoplastic elastomers

Thermoplastic elastomers based on blends of polyolefins with diene rubbers such as butadiene rubber (BR), natural rubber (NR) and styrene–butadiene rubber (SBR) were first described by Coran and Patel [22–24]. These compositions have fairly good initial tensile properties. Moreover, their thermal stability is somewhat better than that of thermoset diene rubbers. Recently, much work has centered on TPEs based upon NR–PP blends. Campbell, Elliott and Wheelans [25] have prepared compositions of partially vulcanized NR. More recently, Payne *et al.* [26] have reported improved mechanical properties of fully vulcanized NR–PP-based TPEs of different hardnesses. They found that unlike thermoset NR, these TPEs have very good resistance to cracking in ozone (Table 6.5). The low temperature

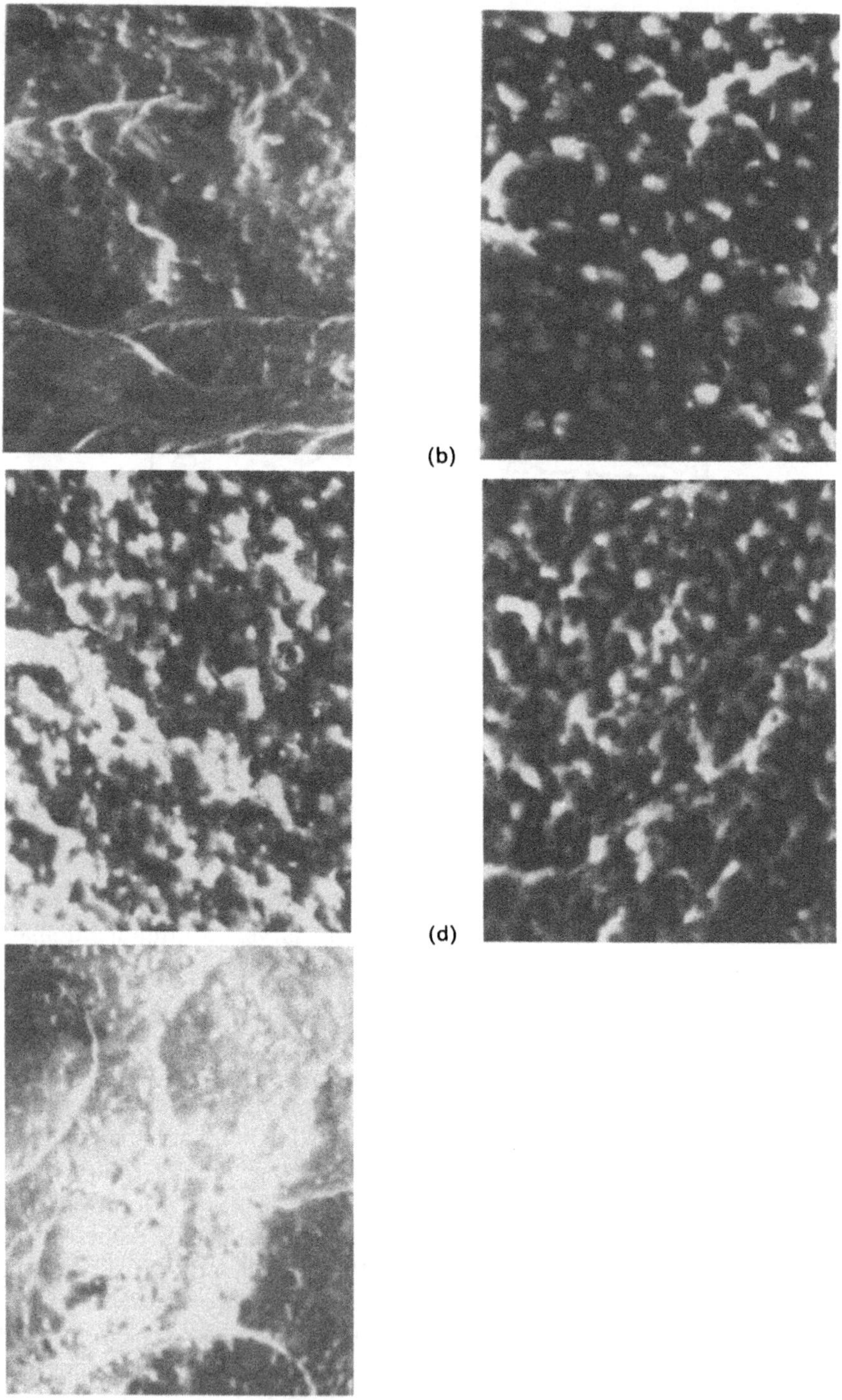

(a) (b) (c) (d) (e)

Figure 6.9 Morphology of EPDM–PP blends. 10 kV, 2.00 kX, 5.00 μm length of bar. (a) 20/80, (b) 40/60, (c) 50/50, (d) 60/40 and (e) 80/20.

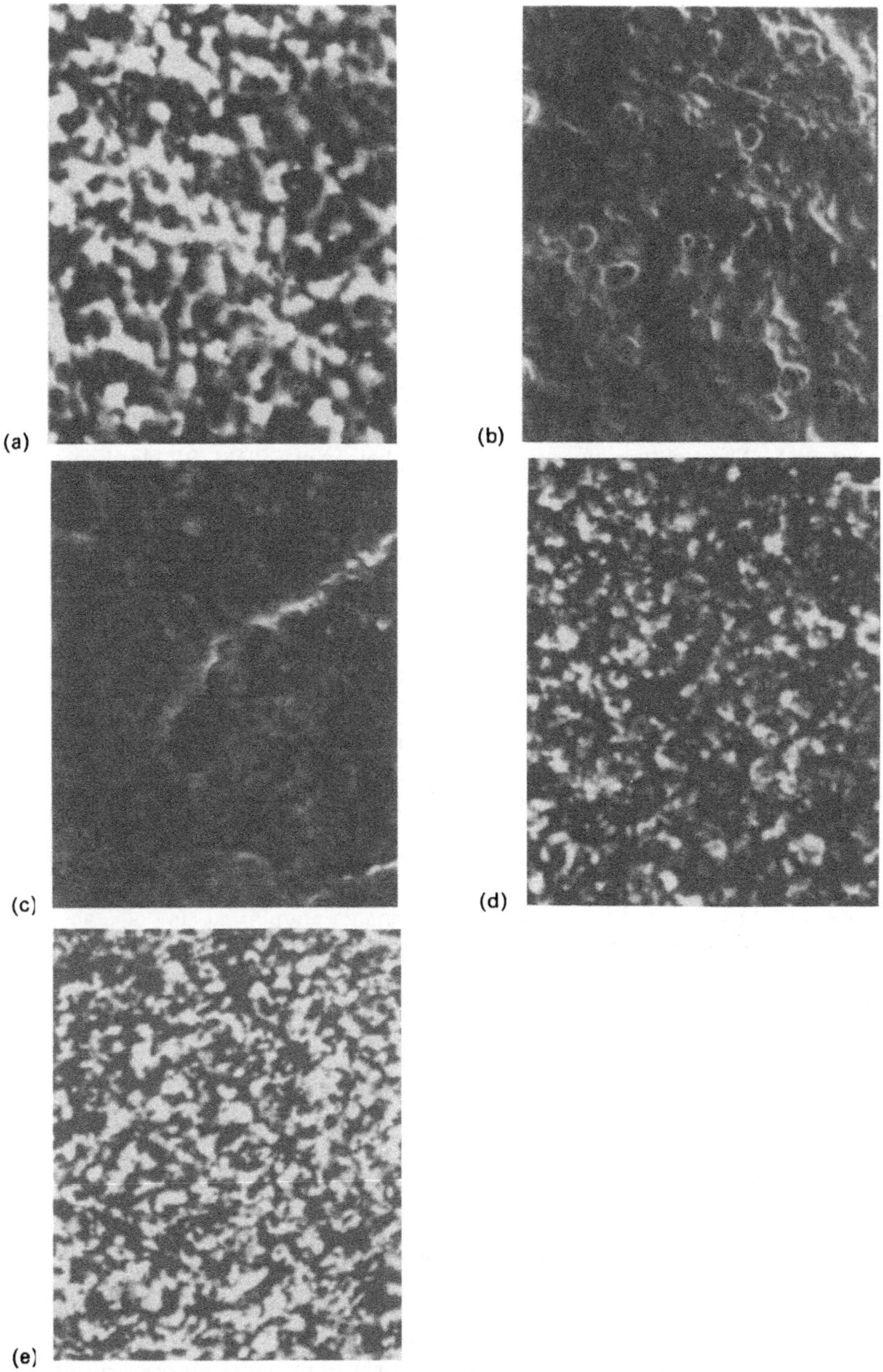

Figure 6.10 Morphology of 60/40 EPDM–PP blend based TPEs. 10 kV, 2.00 kX, 5.00 μm length of bar. (a) uncured blend, (b) partially crosslinked, (c) partially crosslinked, (d) fully crosslinked and (e) fully crosslinked.

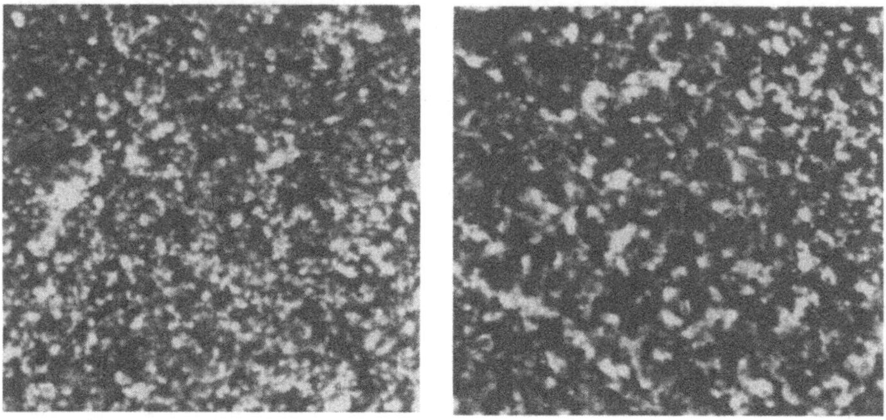

Figure 6.11 Effect of type of crosslink on morphology of EPDM–PP blend based TPEs. 10 kV, 2.00 kX, 5.00 μm length of bar.

Table 6.5 Mechanical properties of different NR–PP-based TPEs

Property	*Method*				
Shore hardness	D2240	60A	70A	90A	50D
Tensile strength (MPa)	D412	5.0	7.6	11.4	20.8
Stress at 100% strain (MPa)	D412	2.1	3.7	6.5	10.5
Ultimate elongation (%)	D412	300	380	400	620
Tension set (%)	D412	10	16	35	50
Tear strength (kN/m)	D624	22	29	65	98
Compression set (22 h)	D395				
23 °C (%)	Method B	24	26	32	45
100 °C (%)		30	32	38	63
Brittle point (°C)	D746	−50	−50	−45	−35
Ozone resistance at 40 °C, 100 ppm of O_3[a]	D518	10	10	10	10
Density (g/cm^3)	D297	1.04	1.04	1.02	0.99

[a] Ozone rating of 10 indicates no cracks after specified time.

brittle point increases with decreases in the natural rubber content. These compositions have fairly good retention of tensile properties in hot air at 100 °C for up to one month (Table 6.6).

6.5.3 Butyl rubber–PP-based thermoplastic elastomers

Butyl rubber (IIR) and halogenated butyl rubbers have very low air and moisture permeability. This makes them well suited for use in tires, and in sporting goods and medical applications. This also provides an opportunity to develop TPEs of low air and moisture permeability.

Table 6.6 Effect of hot air aging on percentage retention of tensile properties of NR–PP-based TPEs

		Aging time (Days at 100 °C)			
		1	7	15	30
TPE hardness	Property				
60A	Tensile strength	99	91	80	40
	Stress at 100% strain	104	65	80	68
	Ultimate elongation	98	110	126	85
70A	Tensile strength	100	87	76	43
	Stress at 100% strain	100	90	86	80
	Ultimate Elongation	98	110	113	56
90A	Tensile strength	103	91	86	66
	Stress at 100% strain	107	103	104	99
	Ultimate elongation	92	93	93	60
50D	Tensile strength	101	95	80	66
	Stress at 100% strain	108	109	102	103
	Ultimate elongation	93	93	91	70

In the early 1960s, Gessler [13] first reported the preparation of toughened (semi-hard to hard – not rubbery) PP compositions by dynamic vulcanization. These compositions were prepared by melt blending a mixture of chlorobutyl rubber, PP and vulcanizing agent, zinc oxide. The work was later greatly expanded by Coran and Patel [27], who were able to produce soft, rubbery thermoplastic compositions, with crosslinked particulate rubber phase (Figure 6.12), the elastomeric properties of which approached those of conventional thermoset rubbers. Since the crosslinked rubber of such a composition is in the dispersed particulate form, one would not expect typical butyl rubber impermeability from these TPE compositions. However, by careful selection of compounding ingredients, it was possible to prepare butyl–PP-based or halobutyl–PP-based TPEs of very low permeability (Table 6.7). The air and water permeabilities of butyl-based dynamic vulcanizates (DVAs) are compared to those of conventional thermosets and of EPDM–PP based TPEs in Figure 6.13.

Recently, Puydak and Hazelton [28] reported the preparation of rubber–rubber–plastic triblend-based TPEs. In most of these compositions two elastomers are used with a plastic phase, preferably PP. One of the elastomers is crosslinked under dynamic vulcanization conditions during molten-state mixing with the second elastomer and a thermoplastic polymer. TEM micrographs of such compositions are shown in Figure 6.14. These TPEs offer some advantages over two-component TPEs (Table 6.8).

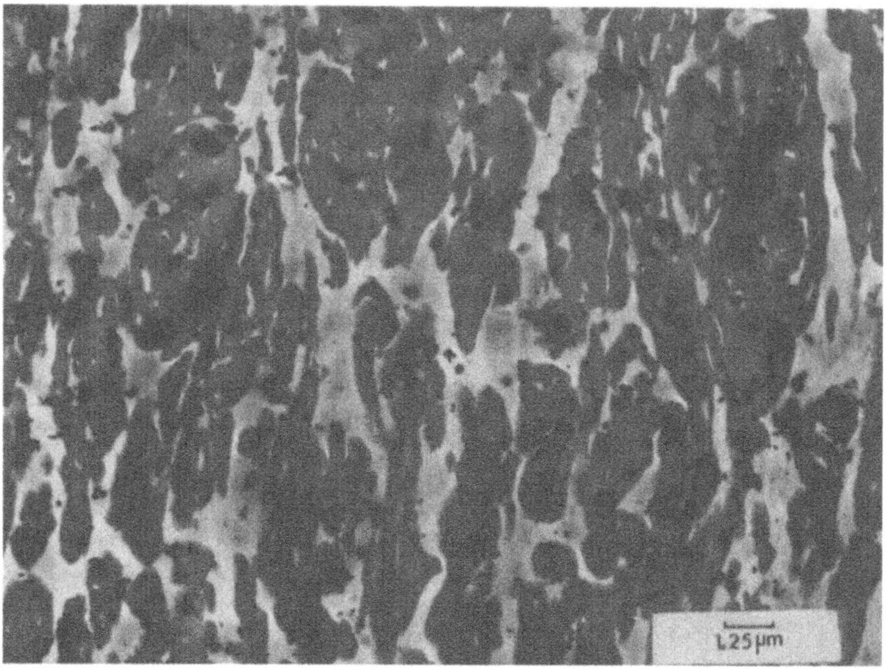

Figure 6.12 TEM micrograph showing morphology of butyl rubber-PP blend based TPE.

Table 6.7 Air permeability of thermoset butyl rubber versus TPEs

Elastomer	*Relative air permeability*[a]
Butyl–PP TPE	1.45
EPDM–PP TPE	4.44
Thermoset butyl (inner liner)	1.00

[a] ASTM D 1434 at 35 °C; sample thickness 0.76 mm.

6.5.4 Thermoplastic elastomers based on PP and other types of rubber

Thermoplastic elastomeric compositions have been prepared by the dynamic vulcanization of PP with a number of types of rubbers in addition to those discussed above. Mechanical properties of compositions of PP with these other types of rubbers, as well as with the types mentioned above, are given in Table 6.9. Each of the compositions contained 60 parts by weight of rubber, 40 parts of PP and a curing or vulcanization ingredient. In each case the curing system was optimized to give the best combination of ultimate tensile strength and ultimate elongation.

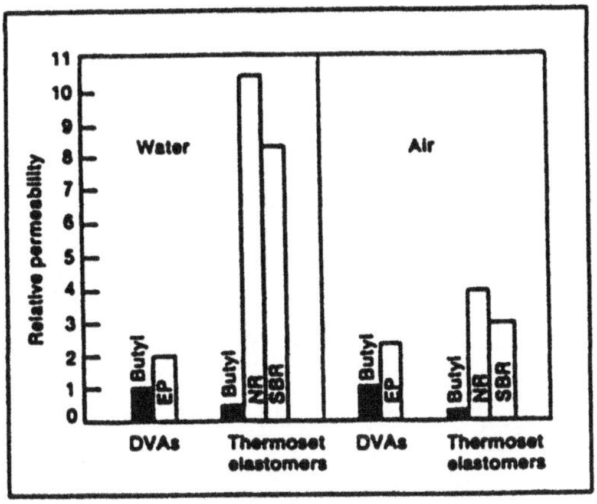

Figure 6.13 Comparison of permeability of thermoset rubbers with TPEs of butyl–PP and EPDM–PP blend based TPEs.

From this data, it is apparent that the strongest and toughest compositions are obtained with rubbers that are almost compatible with PP, e.g. the aliphatic hydrocarbon rubbers.

It should be noted that a systematic study of blends of a large number of different types of rubbers and a large number of different types of plastics by Coran and Patel [29] showed that the best compositions are obtained when there is a good match of surface energy between the rubber and plastic, when the plastic phase is highly crystalline (as in the case of PP), and when the rubber is of the type of high entanglement density. The acrylate rubber (ACM) has the *lowest* entanglement density, and this could be the reason why the properties of the ACM–PP composition of Table 6.9 are so poor.

6.6 THERMOPLASTIC ELASTOMERS FROM NONCOMPATIBLE ELASTOMER AND PP: TECHNOLOGICAL COMPATIBILIZATION OF NBR–POLYOLEFIN BLENDS

Were it not for their gross mutual incompatibility (in the thermodynamic sense), a combination of a polyolefin resin with nitrile butadiene rubber (NBR) might be a good choice of materials from which to prepare oil resistant thermoplastic elastomeric compositions by dynamic vulcanization. Early work with these materials demonstrated only marginal success in obtaining good mechanical properties for such compositions [15, 16, 22]. This was probably due to the large surface energy difference between the two types of polymers. Mutual wetting between the polymers appeared incomplete; relatively large particles of cured rubber dispersed in polyolefin resin formed during mixing and dynamic vulcanization. An approach to technological compatibilization, in addition to dynamic vulcanization, was thus sought by Coran and Patel [8, 30].

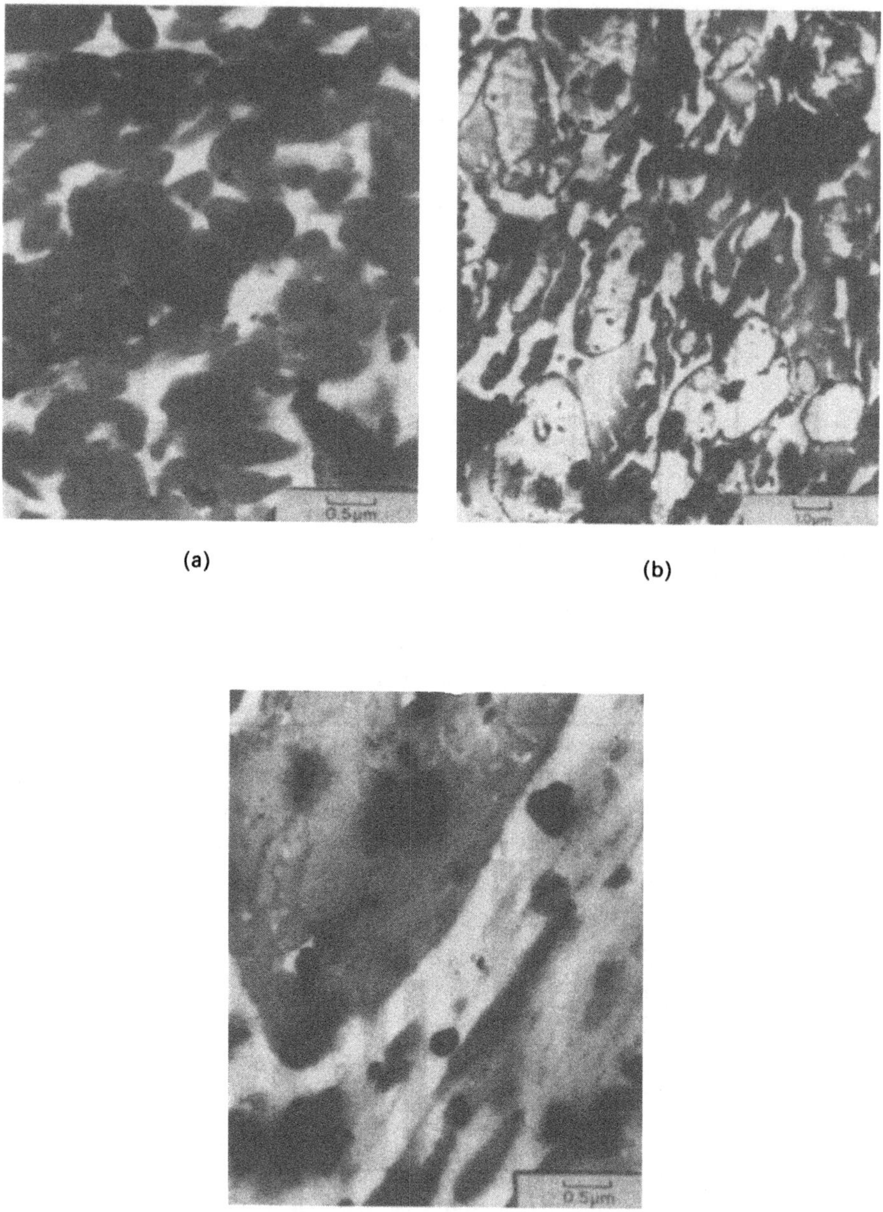

(a) (b) (c)

Figure 6.14 (a) TEM image of elastomeric TPE showing micrometer-sized cured rubber particles in a continuous PP matrix. (b) TEM image of a TPE containing two elastomers, polychloroprene (large, irregular particles outlined in black) and butyl (small dark grey particles). (c) TEM image of a TPE with a crosslinked phase (lighter grey) surrounded by an uncured phase (darker grey) in a continuous polymer phase (white).

Table 6.8 Synergism on combining elastomers in TPEs

TPE property	*polychloroprene (CR)*	*Butyl*	*CR+butyl (1:1)*
Tensile strength (MPa)	4.7	7.1	7.5
Ultimate elongation (%)	170	220	250
Hardness, Shore A	58	60	66
ASTM #3 oil swell (70 h at 100 °C) (%)	21	75	44
Tear strength (kN/m)	12.3	19.3	24.5
Compression set (22 h at 100 °C) (%)	66	47	52

Table 6.9 Tensile properties of different (60/40) rubber–plastic blends

Rubber[a]	*Curative*[b]	*Tensile strength (MPa)*	*Ultimate elongation (%)*	*Tension set (%)*
IIR	P	21.6	380	23
EPDM	S	24.3	530	16
PTPR	P	22.7	210	20
NR	P	26.4	390	24
BR	M	20.8	260	27
SBR	P	21.7	430	30
EVA	O	17.8	320	36
ACM	SO-S	4.0	20	failed
CPE	M-O	12.3	310	55
CR	S	13.0	140	33
NBR	M-O	17.0	200	31

[a] IIR, butyl, PTPR, poly(*trans*-pentenamer) (Bayer); EVA, ethylene–vinyl acetate copolymer rubber; ACM, polyacrylate rubber; CPE, chlorinated polyethylene rubber; NBR, nitrile butadiene rubber.

[b] P, dimethylol phenolic; S, accelerated sulphur; M, bismaleimide; O, organic peroxide, SO-S, soap-sulphur; M-O, bismaleimide-peroxide.

It is now generally accepted that a block copolymer can compatibilize mixtures of the 'parent' homopolymers. The block copolymers act as macromolecular surfactants to promote and stabilize the emulsion of the molten homopolymers [6, 7, 31, 32]. Figure 6.15 is an idealized visualization of this.

It has been found that a dimethylol phenolic compound (such as 'phenolic resin' curative) can be used to technologically compatibilize a mixture of polyolefin and NBR. This compatibilization could be the result of the formation of a block copolymer of the type visualized in Figure 6.15. Such a

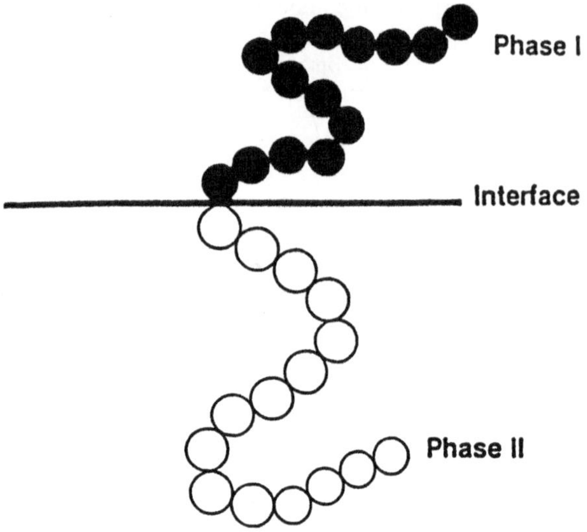

Figure 6.15 Idealized compatibilizing block copolymer.

compatibilizing block copolymer could be formed by the following reaction scheme:

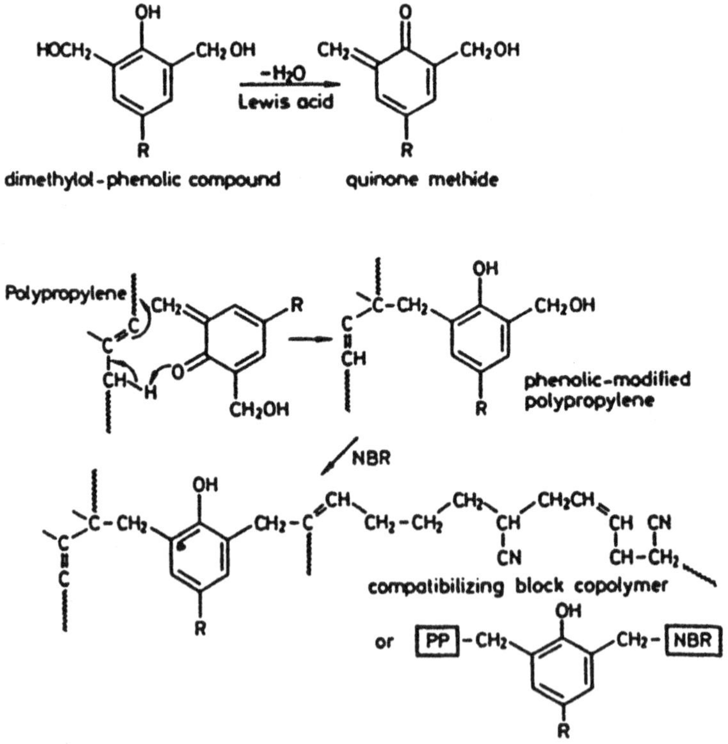

The scheme requires the presence of olefinic unsaturation in the PP molecules. Indeed, to satisfy demands of strict polymerization stoichiometry, there is, on average, one double bond per polyolefin molecule. The scheme is similar to that proposed long ago for the phenolic-resin-curing of diene rubbers [33–35].

In practice, the polyolefin resin is treated with about 1–4 parts of a phenolic curative (e.g. SP 1045) per 100 parts of polyolefin resin (e.g. PP) in the presence of 0.1–0.5 parts of a Lewis acid (e.g. $SnCl_2$) at a temperature of about 180–190 °C. The phenolic-modified polyolefin is then melt-mixed with NBR for a sufficient time for compatibilization to occur, with the formation of a blend of improved homogenization. Then with continuing mixing, curative for the rubber is added. (This can be additional dimethylol phenolic resin curative.) If the NBR contains a small amount (ca. 5%) of an amine-terminated liquid NBR (e.g. Hycar® ATBN 1300X16, B. F. Goodrich Co.), then the properties of the compatibilized blend are even more improved.

The formation of polymer–polymer grafts can be accomplished by a number of other chemical means, in addition to the above use of dimethylol phenolic derivatives. In some cases the results are even better. Such a case is the use of maleic-modified PP to form the block-polymeric compatibilizing agent by reaction with the amine terminated liquid NBR. In this case, polypropylene is modified by the action of either maleic acid or maleic anhydride in the presence of decomposing organic peroxide [36, 37]. During the process, the molecular weight of the PP becomes greatly reduced and the molecules acquire pendant succinic anhydride groups:

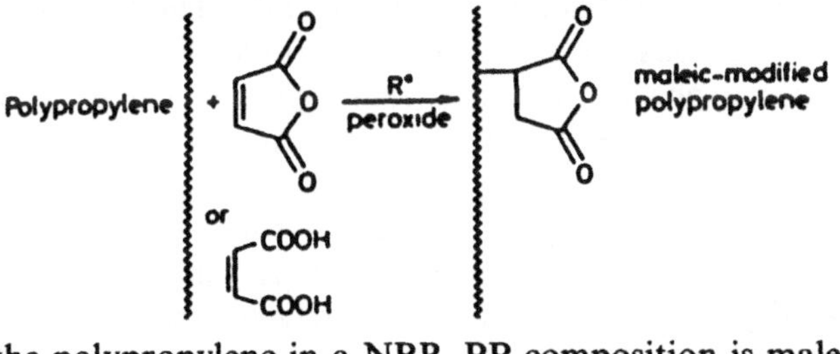

If part of the polypropylene in a NBR–PP composition is maleic-modified, and if part of the NBR is amine-terminated, then a compatibilizing amount of NBR–PP block copolymer forms *in situ* during melt-mixing:

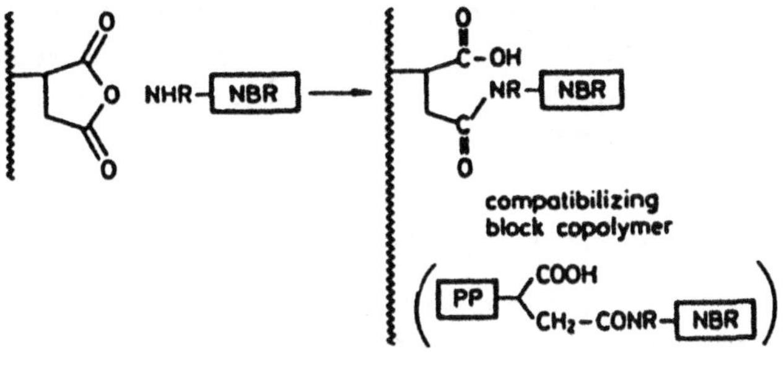

Only a small amount of compatibilizing block copolymer is needed to obtain a substantial improvement in the properties of a blend. The data of Table 6.10 relate to compositions in which 10% of the PP is maleic-modified (by the action of 5 parts of maleic anhydride in the presence of 0.87 parts of Lupersol®-101 peroxide per 100 parts of PP at 180–190 °C). Varying amounts of the NBR are replaced by amine-terminated liquid NBR. Since the maleic-modified polypropylene is generally in stoichiometric excess, it can be assumed that essentially all of the amine-terminated rubber is grafted to some of the PP. After each compatibilized blend was prepared, it was dynamically vulcanized and subjected to the usual treatment for molding and testing.

The data of Table 6.10 indicate that improved blend properties are obtained when as little as 0.16% of the rubber is grafted to PP. Also, after as much as about 2–3% of the rubber is grafted to the polyolefin, additional graft formation gives no further improvement. It should be noted that the mechanical properties of the compatibilized dynamically vulcanized blends of NBR and PP can be about as good as those of dynamically vulcanized EPDM and PP.

The hot oil resistance and brittle point of a compatibilized NBR-PP thermoplastic vulcanizate prepared by dynamic vulcanization is given by Table 6.11 (Stock 1). Although the hot oil resistance is excellent, low temperature performance is somewhat lacking.

It has been found that the low temperature brittle point can be reduced with a minimum sacrifice of hot oil resistance by blending the compatibilized NBR–PP composition with a commercially available thermoplastic elastomer based on PP and vulcanized EPDM. The two thermoplastic elastomer compositions are mutually compatible, since they are both based on a continuous phase of PP. The results given in Table 6.11 indicate that the blend exhibits average mechanical properties but surprisingly better than average hot oil resistance.

The results of this work suggest a practical route to hot-oil-resistant thermoplastic elastomers based on NBR and a polyolefin resin, although these two types of polymer are normally grossly incompatible with each other. The mechanical properties of such blends can approach those of dynamically vulcanized EPDM–PP blends and, in addition, excellent hot oil resistance can be achieved. The NBR-based compositions may not have good enough resistance to low-temperature embrittlement for some applications, but compositions based on EPDM and PP can be blended with the NBR–PP composition to improve the brittle point without causing severe losses of other attributes.

6.7 MULTICOMPONENT THERMOPLASTIC ELASTOMERS

In this section we describe thermoplastic elastomer compositions comprising technologically compatibilized blends of TPEs based on polyamide (PA)

Table 6.10 Effect of graft copolymer concentration on the properties of compatibilized-blend-based TPEs

	1	2	3	4	5	6	7	8
Recipe[a]								
PP[b]	50	45	45	45	45	45	45	45
Maleic-modified PP	–	5	5	5	5	5	5	5
NBR[c]	50	50	49.22	46.88	43.75	37.5	25	–
NBR Masterbatch[d]	–	–	0.78	3.12	6.25	12.5	25	50
SP-1045[e]	3.75	3.75	3.75	3.75	3.75	3.75	3.75	3.75
$SnCl_2 \cdot 2H_20$	0.50	0.50	0.50	0.50	0.50	0.50	0.50	0.50
ATBN[f] as % of NBR (% rubber grafted to PP)	0.00	0.00	0.16	0.62	1.25	2.5	5	10
Properties								
Tensile strength $^{\sigma}_{B}$ (MPa)	8.8	12.0	12.4	15.2	22.0	25.5	25.7	26.7
Stress at 100% strain (MPa)	–	12.0	12.1	12.0	12.3	12.3	12.5	12.9
Young's modulus E (MPa)	209	200	212	223	185	188	184	237
Ultimate elongation ε_B (%)	19	110	170	290	400	440	430	540
Tension set ε_s (%)	–	–	45	40	40	40	42	45
True stress at break σ_B^* (MPa)	10	25	33	59	110	138	136	171
Breaking Energy[g] (J/cm^3)	1.2	11	18.4	34.6	54.9	64.1	61.7	86.5
Improvement in breaking energy due to compatibilization (%)	–	–	67	215	399	483	461	686

[a] Parts by weight.
[b] Profax®6723. ATBN is Hycar® ATBN 1300 × 16 amine-terminated liquid nitrile rubber.
[c] Hycar®1092-80 nitrile rubber.
[d] 90% by weight of Hycar®1092-80, 10% by weight of Hycar® ATBN 1300 × 16 liquid nitrile rubber.
[e] A methylolphenolic vulcanizing agent.
[f] A quantitative reaction is assumed between ATBN and maleic-modified PP (which is assumed to be present in excess).
[g] Breaking energy values were from stress–strain curves.

Table 6.11 Properties of compatibilized NBR–PP TPE composition blended with EPDM–PP TPE composition

	1	2	3
NBR–PP composition (wt %)[a]	100	50	0
EPDM–PP composition (wt %)[b]	0	50	100
Properties			
Tensile strength (MPa)	22.6	15.9	8.6
Stress at 100% strain (MPa)	11.2	7.9	4.4
Ultimate elongation (%)	585	510	415
Hardness, A scale	93	87	68
Tension set (%)	48	23	10
True stress at break (MPa)	155	97	44.3
ASTM No. 3 oil volume swelling (%)[c]	22	32.5	62.5
Brittle point (°C)	−24	−47	< −60

[a] Recipe for the NBR–PP composition is: Profax®6723 polypropylene, 45 (parts by weight); maleic-modified polypropylene, 5; Hycar®1092-80 NBR/Hycar®ATBN 1300 × 16 liquid NBR (90:10), 50; SP-1045 phenolic curative, 3.75; $SnCl_2 \cdot 2H_2O$, 0.50; Naugard™ 495 stabilizer (added after vulcanization), 1.0.

[b] Commercially available Santoprene®201-73 thermoplastic elastomer (Advanced Elastomers Systems).

[c] Hot-oil swelling, 70 h at 100 °C.

and PP. Both the PA- and the PP-based TPEs were first prepared by dynamic vulcanization. The TPE blend compositions were technologically compatibilized by the presence of chemically modified PP, which presumably reacted with small portions of the PA to form compatibilizing amounts of PA–PP graft-linked copolymer [30, 38]. Components of the blends of thermoplastic vulcanizates are described below.

EPDM–PP TPE: a 50/50 blend of EPDM and PP was dynamically vulcanized by 5 parts (per 100 parts of combined polymers) of SP 1045 dimethylol phenolic curative in the presence of 1 part of $SnCl_2 \cdot 2H_2O$ at 180–190 °C.

NBR–PA TPE: a 65/35 blend of NBR and PA 6,6-6 copolymer (melting point, 213 °C) was treated with 1.3 parts (per 100 parts of combined polymers) of SP 1045 phenolic curative during its mixing at 215 °C. (Flectol®H antidegradant was used at a level of 2 parts per 100 parts of polymer.)

Polyurethane Rubber–PA TPE: a 50/50 blend of a millable, vulcanizable polyurethane rubber (Adiprene®C) and PA 6,6-6,6-10 (melting point 163 °C, Zytel®63) was dynamically vulcanized at 180 °C by the action of 1.0 part of *m*-phenylenebismaleimide in the presence of 0.5 parts of L-101 organic peroxide per 100 parts of combined polymers. The product was designated PU–PA TPE.

Epichlorohydrin Rubber–Nylon TPE: a 50/50 blend of Hydrin 400 epichlorohydrin rubber (ECO) and PA 6,6-6,6-10 (melting point 163 °C,

Zytel®63) was dynamically vulcanized at 170–180 °C by the action of 1.67 parts of zinc stearate, 1.0 part of bisbenzothiazole disulfide (MBTS), and 0.4 part of sulfur per 100 parts of combined polymers, in the presence of 1.0 part of Flectol® H antidegradant. The product was designated ECO–PA TPE.

Maleic acid-modified polypropylene: this functionalized polymer was prepared by melt-mixing 100 parts of polypropylene with 5 parts of maleic acid in the presence of 2.0 parts of L-101 peroxide (added after the acid and polymer were well mixed) at 180 °C for about 3 minutes. The product was designated MA–mod. PP.

Mixtures of TPEs: blends of the various thermoplastic elastomers (with and without the chemically modified, compatibilizing, functionalized polypropylene) were prepared by melt-mixing the ingredients in a Brabender or Haake Rheomix internal mixer at about 10 °C above the melting point of the PA used in each case.

Properties of molded sheets of the mixtures of thermoplastic elastomers are given in Table 6.12. These test results indicate that PA-based TPEs containing a variety of different types of rubbers can be used in the compatibilized blends. A substantial improvement in properties is obtained in each case. (Note the comparison between the control stocks containing added unmodified polypropylene (odd-numbered stocks) and the experimental stocks containing the maleic-modified polypropylene (even-numbered stocks).)

It is believed that the maleic acid-modified PP molecules contain pendant succinic anhydride groups which react with amine groups in the nylon to form compatibilizing amounts of a graft-linked PA–PP block copolymer [36]. It should be noted that a number of other chemical modifications of PP have also given results similar to those obtained with maleic acid-modified PP [30, 38].

From this work, one can conclude that compositions that have excellent mechanical properties can be prepared by melt-mixing thermoplastic vulcanizates (TPV) (which had been prepared previously by dynamic vulcanization). Excellent mixed-TPV compositions can be obtained even though the rubbers and plastics are mutually grossly incompatible with respect to thermodynamic considerations. In such cases, however, it appears to be necessary for a compatibilizing agent to be present in the mixture to promote the interaction between the thermoplastic materials.

The rubber associated with one of the thermoplastic components can differ greatly from the rubber associated with another thermoplastic component. Thus, a composition can be produced which has good mechanical properties and which can contain both different thermoplastic resins and different rubbers.

Table 6.12 Properties of compatibilized PP–PA TPE compositions

	1	2	3	4	5	6
Recipe						
Dynamically vulcanized EPDM–PP composition	50	50	50	50	50	50
Dynamically vulcanized NBR–PA composition	50	50	–	–	–	–
Dynamically vulcanized PU–PA composition	–	–	50	50	–	–
Dynamically vulcanized epichlorohydrin rubber–PA composition	–	–	–	–	50	50
Polypropylene	10	–	10	–	10	–
Maleic-modified polypropylene	–	10	–	10	–	10
Properties						
Tensile strength (MPa)	16.1	26.9	9.2	19.6	11.9	22.0
Stress at 100% strain (MPa)	14.0	16.5	–	11.8	9.3	12.0
Young's modulus (MPa)	132	147	173	157	206	199
Ultimate elongation (%)	170	310	73	270	230	340
Tension stress (%)	40	39	–	52	40	50
True stress at break (MPa)	43	110	16	73	39	97

6.8 APPLICATIONS

The rubber–plastic blends discussed in this chapter are generally intended for use as TPEs. These are materials which have many of the properties of conventional vulcanized (thermoset) rubbers, but are processable and can be shaped into parts by the rapid techniques used for thermoplastic materials.

Thermoplastic processing is far more economically attractive than traditional multi-step rubber processing. In the case of conventional thermoset rubber processing, the producer of rubber articles purchases gum rubber, fillers, extender oils or plasticizers, curatives, antidegradants, etc.; these ingredients must then be mixed and uniformly dispersed; after a stock is mixed, it is then shaped by extrusion, calendering, etc.; the crudely shaped preform is then vulcanized in its final shape in a mold contained by a press and heated for vulcanization, which can take a long period of time. Mold flash or overflow, as well as rejected parts, are not reprocessable without expensive reclaiming to break down the crosslinked rubber network.

On the other hand, a part produced from a thermoplastic elastomer is shaped or molded into its final shape in a single step. Also mold flash and rejected parts can simply be ground and reused. Detailed comparisons have shown time and again that thermoplastic processing is more economical than thermoset rubber processing [4, 39].

The lower cost of thermoplastic processing is the reason for the development of the TPEs. However, failure in the achievement of truly rubber-like properties, in many cases, has impeded the acceptance of TPE technology.

The application of TPE technology has been reviewed briefly by O'Connor and Fath [39]. The first thermoplastic elastomers were polyurethanes developed in the 1930s (in Germany) and introduced commercially in the United States in the late 1950s [40]. These materials gained rapid acceptance because of their high-performance characteristics: solvent resistance, abrasion resistance, high tear strength, good load bearing capacity and low-temperature flexibility. Continued growth in the use of these materials at the expense of vulcanized rubber usage is unlikely because their high price cannot be offset by the reduction in processing costs resulting from thermoplastic processing [41].

Styrenic triblock (polystyrene end segment–rubbery center segment–polystyrene end segment) copolymers [42, 43] were the second type of thermoplastic elastomer to be commercialized (in the early 1960s). These low-cost elastomers have achieved the highest volume of use, thus far, for thermoplastic elastomers. This was because of their acceptance in the footwear and adhesives markets [41]. However, further significant growth in the use of the styrenic thermoplastic elastomers depends on acceptance in other markets.

Copolyester thermoplastic elastomers (poly[oxy-1,4-butylene]–poly-1,4-butylene terephthalate alternating block copolymers) were then introduced

Table 6.13 Properties of various types of elastomer compositions

Property	*Partially vulcanized EPDM–PP blend*[a]	*Completely vulcanized EPDM–PP blend*[b]	*CR vulcanizate*	*Ester–ether copolymer TPE*[c]
Hardness, A scale	77	80	80	92
Tensile strength (MPa)	6.6	9.7	9.7	25.5
Ultimate elongation (%)	200	400	400	450
ASTM No. 3 oil swell (70 h at 100 °C) (vol%)	–[e]	50	35	30
Compression set (ASTM method B, 22 h at 100 °C)	70	39	35	33
Upper use temperature (°C)	100	125	110	125
Type of processing[d]	TP	TP	CV	TP

[a] TPR-1700 (Uniroyal).
[b] Santoprene® (Advanced Elastomer Systems).
[c] Hytrel® (du Pont).
[d] TP, – thermoplastic processing; CV – conventional vulcanization.
[e] Sample disintegrated.

in the early 1970s. They are high-performance materials with excellent solvent resistance, abrasion resistance, ultimate properties, fatigue resistance and very low hysteresis. Their acceptance has been limited only by their relatively high price and high hardness [4].

Thermoplastic olefinics (TPOs) were also introduced in the early 1970s. These are blends of EPDM and polyolefin plastic (usually PP) such as those discussed earlier in this chapter. The rubber was slightly or not at all crosslinked [1]. These blend compositions replaced thermoplastic urethane elastomers and even vulcanized rubbers in some exterior automotive components. They replaced polyvinylchloride (PVC) in wire and cable applications, and replaced thermoset rubber in mechanical-goods applications, though to a limited extent.

The polyurethane and copolyester types of thermoplastic elastomers have good performance properties but they are relatively expensive materials. The styrenics and TPOs are relatively inexpensive but they are poor in solvent and fluid resistance. They have poor compression set and tensile properties even at moderately elevated temperatures.

The more recently commercialized compositions based on PP and *completely* vulcanized EPDM (the thermoplastic vulcanizates prepared by dynamic vulcanization, described earlier) have many of the excellent properties of the polyurethane and copolyester type thermoplastic elastomers and even improved set and fatigue properties. Comparisons between the properties of commercial grades of completely vulcanized EPDM-PP TPE materials and those of conventional vulcanized specialty rubbers, such as

CR, EPDM and chlorosulfonated PE (CSM) have been made by O'Connor and Fath [39]. The comparison of selected properties of various thermoplastic elastomers and those of conventional elastomers are given in Table 6.13.

6.9 PROCESSING/FABRICATION TECHNOLOGY

The processing of a rubber–plastic blend composition into a finished part is a function of the melt rheology of the composition, and the melt rheology is a function of temperature and shear rate. Processing is also a function of the strength of the molten material under the strain due to its processing, i.e., a function of its resistance to melt fracture. Melt fracture, under the conditions of extrusion, can give rise to very poor surface textures or even functionally useless parts.

The melt rheology of a rubber–plastic blend composition is related to that of the plastic material. This is illustrated by the idealization given by Figure 6.16. At high shear rates the viscosity–shear rate profiles are similar for the blend and the plastic material, *per se*. However, at very low shear rates, the viscosity of the blend can be very high. In the case of highly rubber-loaded thermoplastic vulcanizates (prepared by dynamic vulcanization) the viscosity can approach infinity where the shear rate is zero. Thus, under the conditions of melt extrusion, the molten material undergoes rapid flow in the die. Then, as the material passes out of the die, the rate of deformation drops to zero and, since the viscosity approaches infinity (presumably due to cured rubber particle–particle interference), little or no die swell is observed. The dimensions of extrusion profiles of such materials are, thus easily controlled.

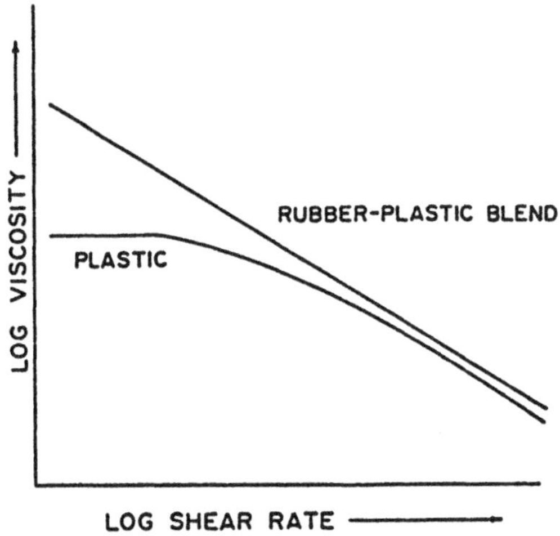

Figure 6.16 The relationship between viscosity and shear rate for a plastic and plastic–rubber blend.

From Figure 6.16, it appears that rubber–plastic blends are highly shear-rate sensitive with respect to melt viscosity. The temperature sensitivity of the viscosity of a thermoplastic elastomeric (highly rubber-loaded) rubber–plastic blend is illustrated by Figure 6.17, which relates to an elastomeric EPDM–PP blend. The viscosity of this type of blend is relatively insensitive to changes in temperature. (In the case of another composition based on another type of thermoplastic phase material, the viscosity may indeed be more temperature-sensitive.) Figures 6.16 and 6.17 indicate that, in processing certain rubber–plastic blends by flow techniques, such as extrusion and injection molding, higher shear rates should be used instead of the common practice of increasing the temperature to facilitate adequate flow.

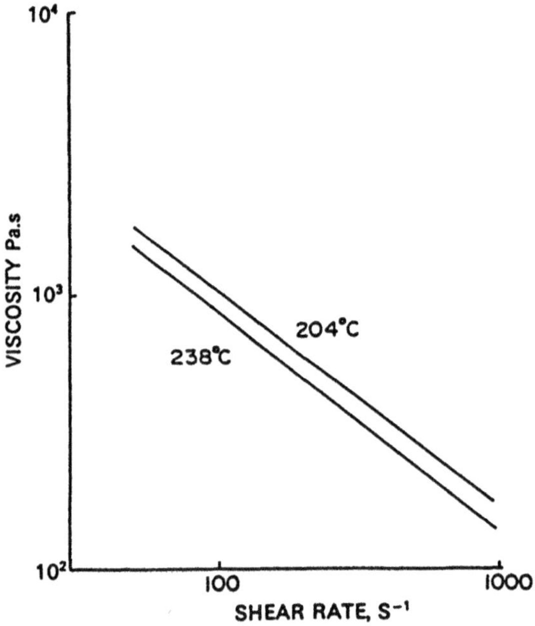

Figure 6.17 The effect of temperature on the viscosity–shear rate relationship for a typical EPDM–PP-based TPE.

As O'Connor and Fath [39] have pointed out, the high melt viscosity of these products can be advantageous in processing. The high viscosity can provide high melt integrity or 'green strength' and permit the retention of shapes of parts produced by extrusion or blow molding. The high melt viscosity and low die swell are also helpful in calendering sheet and film products. For injection-molded parts, fast injection rates (under high pressure) give lower viscosities due to the high shear rate. This facilitates rapid and complete mold filling. Then, after the mold is filled, the viscosity increases greatly due to shear rate reduction (to zero). This increased viscosity, which can approach infinity, enables more rapid extraction of the part from the mold. The overall effect is a faster injection molding cycle. In addition, the low temperature sensitivity of the viscosity of such a composition gives a broad temperature window for processing. Typical injection molding and extrusion conditions for an EPDM–PP polypropylene blend are given in Tables 6.14 and 6.15.

Table 6.14 Injection-molding conditions for EPDM–PP-based TPE blends

Rear-zone barrel temperature (°C)	180–220
Center-zone barrel temperature (°C)	205–220
Nozzle temperature (°C)	205–220
Mold temperature (°C)	20–65
Injection pressure (MPa)	35–140
Hold pressure (MPa)	30–110
Back pressure (MPa)	0.7–3.5
Screw speed (rpm)	25–75
Injection speed	Moderate to fast
Injection time (s)	5–25
Hold time (s)	15–75
Total cycle time (s)	20–100

Table 6.15 Extrusion conditions for EPDM–PP-based elastomeric blends

Rear-zone barrel temperature (°C)	175–210
Center-zone barrel temperature (°C)	175–210
Front-zone temperature (°C)	190–220
Adapter temperature (°C)	200–225
Die temperature (°C)	205–225
Melt temperature (°C)	205–235
Screw speed (rpm)	10–150

In addition to the above (injection molding, calendering, extrusion, and blow molding), foaming, thermoforming, and compression molding of olefinic rubber–plastic blends have been reviewed [1]. However, it should be noted that processing conditions vary widely with equipment mold designs, specific blend compositions, etc. The best conditions for the production of a given part in a given factory must be found by experimentation.

6.10 END USE APPLICATIONS

A large number of TPE rubber–plastic blends, having good properties, can now be prepared. However, this was not thought possible until relatively recently. (Because only very few polymer–polymer combinations are thermodynamically compatible, it was thought that only a few rubber–plastic blend combinations could be useful.) As stated earlier in this chapter, new techniques such as dynamic vulcanization and *technological* compatibilization (compatibilization by block copolymer addition or *in situ* formation) have, only recently, expanded the number of rubber–plastic combinations that can produce useful blends. As a result of this, it is probable that only the beginning of the list of commercial end-use applications is presently visible, and that list of TPE blend applications is largely limited to rubber–PP type TPEs.

Some mention should be made here of NBR–PVC blends. When NBR is the major component, the compositions are generally handled as conventional vulcanizates in which the plastic phase gives improved ozone resistance, etc. Compositions containing about 50% or more of the PVC behave as plasticized PVCs in which the plasticizers are nonmigratory [6, 7]. Thus the thermoplastic NBR–PVC blends do not have truly rubber-like properties since the glass transition temperature of the largely non-crystalline plastic phase is only about 85 °C. At relatively low temperatures, therefore, this composition loses all of its mechanical integrity. Thus, we will consider mainly the end-use applications of the olefinic types.

The end use of an elastomer depends largely on its performance-related properties. Rubbers are generally classified with respect to two of these end-use related properties (ASTM D2000), namely dry heat resistance and resistance to swelling in hot oil. This classification is illustrated by Figure 6.18. This figure gives the positions of the thermoplastic elastomers based on PP blends with respect to heat resistance and hot oil resistance in relation to other elastomers.

6.10.1 Uses of blends containing unvulcanized or only slightly crosslinked olefinic rubber

A major application of EPDM–PP blends of this type has been in exterior automotive body parts such as filler panels, bumper covers, fender extensions, lower fascias, flexible front and rear panels, corner panels, and sight shields [1]. The blends have replaced vulcanized EPDM rubber and thermoplastic polyurethane elastomers. Much of this market has been styling-related and has been subject to the whims of designer/consumer tastes.

Other, more functional exterior automotive applications such as gaskets and weather stripping have been tried. Specific examples include door guards, body and bumper strips, hood gaskets, trunk gaskets, taillight gaskets, and windshield gaskets. One reason for these applications is the excellent weathering resistance of these materials. Interior automotive parts that have been made with the partially vulcanized or unvulcanized EPDM–

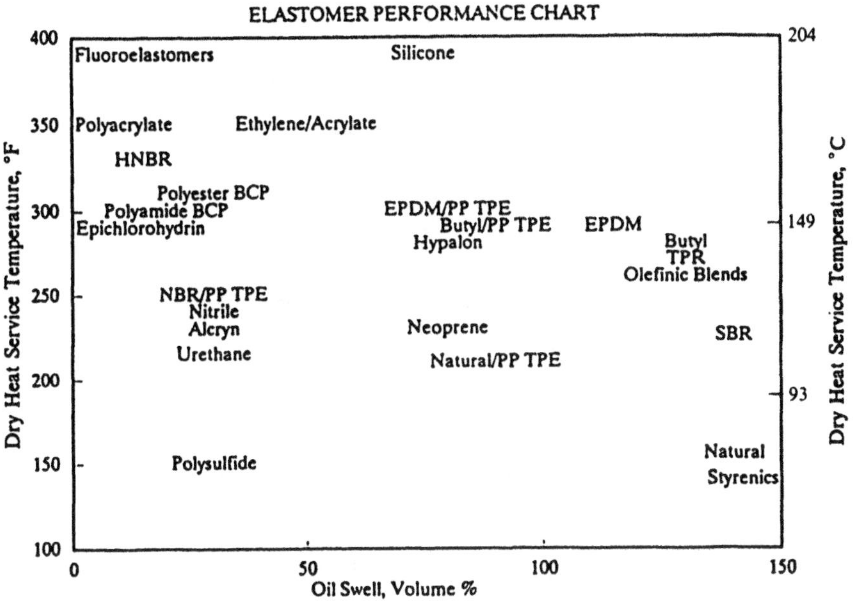

Figure 6.18 Elastomer performance classification according to hot oil and hot air resistance.

PP blends are steering wheels, connector strips, grommets, seals, bushings, seat belt housings, horn pads, flexible trim and certain decorative parts. under-the-hood applications have included sparkplug boots, electrical connectors, hose, and tubing, all of which must be stable against the effects of long-term elevated temperatures.

Wire and cable applications are a large potential market for these blends. They have been used mainly as insulation and jacketing materials. They are competitive with crosslinked polyethylene, which is widely used in power and communication cable. The good resistance to thermal-oxidative degradation, as well as excellent electrical insulating properties, are driving forces suggesting the wire and cable (and other electrical insulation-type) applications.

6.10.2 Uses of thermoplastic vulcanizates containing completely vulcanized rubber particles

The end-use applications of EPDM–PP blends have been greatly expanded by vulcanizing the rubber phase to a high state of cure. This is done commercially by dynamic vulcanization. The potential of this type of material has been expanded because of the improvements (in comparison to those discussed above) with respect to compression set (Figure 6.19),

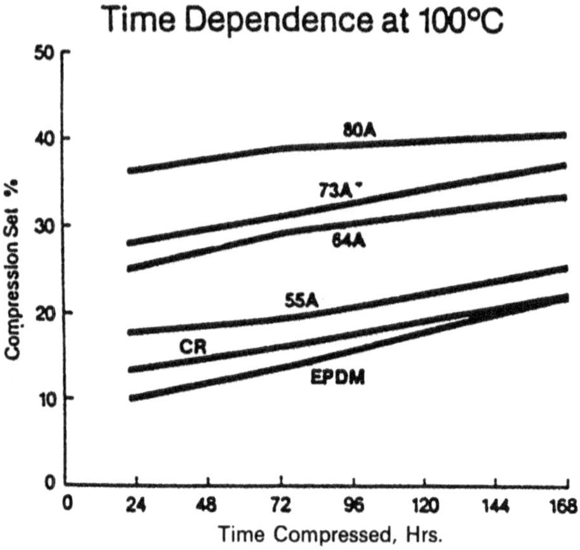

Figure 6.19 Compression set of EPDM–PP-based TPEs of different hardnesses.

ultimate mechanical properties (Figure 6.20), fatigue resistance (Figure 6.21), resistance to attack by fluids (Figure 6.20), high-temperature mechanical integrity, stability of phase morphology in the melt (during processing), melt strength, and thermoplastic fabricability [39, 44, 45]. Thus, these materials are more rubber-like with respect to their performance characteristics, yet they are also more rapidly fabricable as thermoplastics.

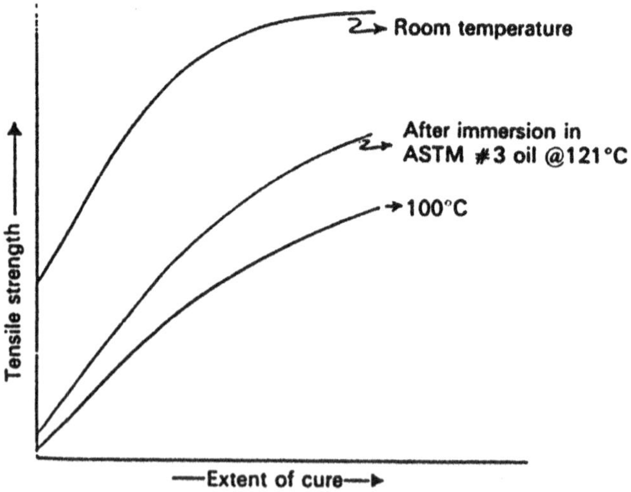

Figure 6.20 Effect of cure level on tensile properties in different environments.

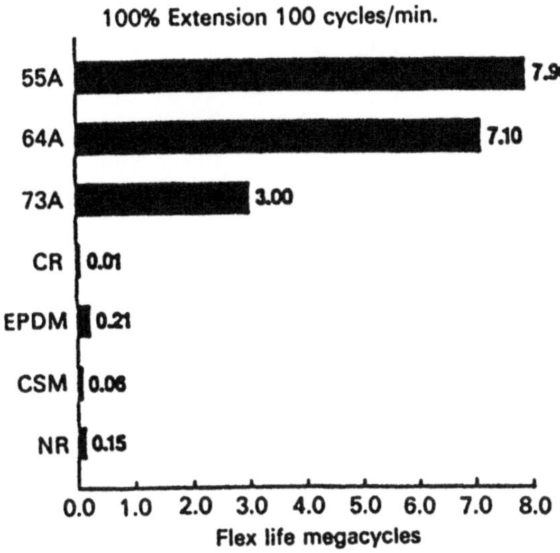

Figure 6.21 Comparison of flex life of EPDM–PP-based TPEs with that of various conventional thermoset elastomers.

Potential and proven applications of these more recently developed materials include:

1. Mechanical rubber goods applications: caster wheels, convoluted bellows, flexible diaphragms, gaskets, seals, extruded profiles, tubing, mounts, bumpers, housings, glazing seals, valves, shields, suction cups, torque couplings, vibration isolators, plugs, connectors, caps, rollers, oil-well injection lines, handles, and grips.
2. Under-the-hood automotive applications: air conditioning hose cover, fuel line hose cover, vacuum tubing, vacuum connectors, body plugs, seals, bushings, grommets, electrical components, convoluted bellows, steering gear boots, emission tubing, protective sleeves, shock isolators, and air ducts.
3. Industrial hose applications: hydraulic (wire braid), agricultural spray, paint spray, plant air–water, industrial tubing, and mine hose.
4. Electrical applications: plugs, strain relief, wire and cable insulation and jacketing, bushings, enclosures, connectors and terminal ends.

6.10.3 Emerging applications

This chapter is very likely to be out of by the time it is published. This is because of the level of product-development/introduction activity presently sustained by the industry. Efforts are directed towards the development of TPE blend compositions that are more resistant to hot oil, compositions that are more resistant to higher temperatures, and, eventually, compositions that are more resistant to both hot oil and higher temperatures.

6.11 APPENDIX: SYMBOLS AND ABBREVIATIONS

ACM	Acrylate rubber
BR	Butadiene rubber
CR	Polychloroprene
DVA	Dynamic vulcanisaty
ECO	Epichlorohydrin rubber
EPDM	Ethylene–propylene diene terpolymer
IIR	Butyl (isoprene-isobutylene) rubber
MBTS	Bisbenzothiazole disulfide
NBR	Nitrite (butadiene) rubber
NR	Natural rubber
PA	Polyamide
PE	Polyethylene
phr	Parts per 100 parts of rubber
PP	Polypropylene
PU	Polyurethane
PVC	Polyvinyl chloride
SBR	Styrene butadiene rubber
TPE	Thermoplastic elastomer
TPO	Thermoplastic olefinic
TPV	Thermoplastic vulcanizate

6.12 REFERENCES

1. Morris, H. L. (1979) in *Handbook of Thermoplastic Elastomers* (ed. B. M. Walker), Van Nostrand Reinhold, New York, p. 5.
2. Kresge, E. N. (1978) in *Polymer Blends*, Vol. 2 (eds D. R. Paul and S. Newman), Academic Press, New York, p. 293.
3. Kresge, E. N. (1984) *Journal of Applied Polymer Science: Applied Polymer Symposium*, **39**, 37.
4. O'Connor, G. E. and Fath, M. A. (1981) *Rubber World*, December.
5. Paul, D. R. and Barlow, J. W. (1979) in *Multiphase Polymers* (eds S. L. Cooper and G. M. Estes), American Chemical Society, Washington, DC, p. 315.
6. Olabisi, O. and Farnam, A. G. (1979) in *Multiphase Polymers*, (eds S. L. Cooper and G. M. Estes), American Chemical Society, Washington, DC, p. 559.
7. Olabisi, O., Robeson, L. M. and Shaw, M. T. (1979) *Polymer–Polymer Miscibility*, Academic Press, New York, pp. 277, 321.
8. Coran, A. Y. and Patel, R. (1983) *Rubber Chemistry and Technology*, **56**, 1045.
9. Stehling, F. C., Huff, T., Speed, C. S. and Wissler, G. (1981) *Journal of Applied Polymer Science*, **26**, 2693.
10. Dunn, J. R. (1976) *Rubber Chemistry and Technology*, **49**, 978.
11. Corish, P. J. and Powell, B. D. (1974) *Rubber Chemistry and Technology*, **47**, 481.
12. Bucknall, C. B. (1977) *Toughened Plastics*, Applied Science Publishers, London.
13. Gessler, A. M. (1962) US Patent 3 037 954, 5 June.
14. Fischer, W. K. (1973) US Patent 3 758 643, 11 September.
15. Coran, A. Y., Das, B. and Patel, R. P. (1978) US Patent 4 130 535, 19 December.

16. Coran, A. Y. and Patel, R. (1980) *Rubber Chemistry and Technology*, **53**, 141.
17. Gesner, B. D. (1969) *Encyclopedia of Polymer Science and Technology*, Vol. 10 (eds H. F. Mark and N. G. Gaylord), Wiley. Interscience, New York, p. 694.
18. Flory, P. J. (1953) *Principles of Polymer Chemistry*, Cornell University Press, Ithaca, NY, pp. 568, 576.
19. Nielson, L. E. (1951) *Review of Scientific Instruments*, **22**, 690.
20. Abdou-Sabet, A. and Fath, M. A. (1982) US Patent 4 311 628, 19 January.
21. Abdou-Sabet, A. and Patel, R. (1991). *Rubber Chemistry and Technology*, **64**, 769.
22. Coran, A. Y. and Patel, R. (1978) US Patent 4 104 210, 1 August.
23. Coran, A. Y. and Patel, R. (1980) US Patent 4 183 876, 15 January.
24. Coran, A. Y. and Patel, R. (1981) US Patent 4 271 049, 2 June.
25. Campbell, D. S., Elliott, D. J. and Wheelans, M. A. (1978) *Natural Rubber Technology*, **9**, 21.
26. Payne, M. P., Wang, D. S. T., Patel, R. and Sasa, M. M. (1990) The expanding technology of thermoplastic elastomers, presented at the ACS Rubber Division meeting, 10–12, October Washington, DC.
27. Coran, A. Y. and Patel, R. (1978) US Patent 4 130 534, 19 December.
28. Puydak R. C. and Hazelton, D. R. (1988) *Plastics Engineering*, 37–9.
29. Coran, A. Y. and Patel, R. (1981) *Rubber Chemistry and Technology*, **54**, 892.
30. Coran, A. Y. and Patel, R. (1982) US Patent 4 355 139, 19 October.
31. Paul, D. R. (1978) *Polymer Blends*, Vol. 2 (eds D. R. Paul and S. Newman), Academic Press, New York, p. 35.
32. Gaylord, N. G. (1975) *Compatibilization Concepts in Polymer Applications* American Chemical Society, Washington, DC, p. 76.
33. Giller, A. (1966) *Kautschule Gummi und Kunststoffe*, **19**, 188.
34. Thelamon, C. (1963). *Rubber Chemistry and Technology*, **36**, 268.
35. Van der Meer, S. (1943) *Revue General du Caoutchouc et des Plastiques*, **20**, 230.
36. Ide, F. and Hasegawa, A. (1974) *Journal of Applied Polymer Science*, **18**, 963.
37. Minoura, M., Veda, M., Mizunuma, S. and Oba, M. (1969) *Journal of Applied Polymer Science*, **13**, 1625.
38. Coran, A. Y., Patel, R. and Williams-Headd, D. (1985) *Rubber Chemistry and Technology*, **58**, 1014.
39. O'Connor, G. E. and Fath, M. A. (1982) *Rubber World*, January.
40. Schollengerger, C. S., Scott, H. and Moore, G. R. (1958) *Rubber World*, **137**, 549.
41. Auchter, J. F. (1981) Outlook for thermoplastic elastomers in the 1980s, presented the ACS Rubber Division meeting, October, Washington, DC.
42. Holden, G. and Milkovich, R. (1965) US Patent 3 265 766.
43. Zelinski, R. P. (1966) US Patent 3 287 333.
44. Abdou-Sabet, A. (1987) Elastomeric alloys – new TPE, presented at the Institution of Mechanical Engineers, London, 26 March 1987.
45. Rader, C. P. and Kear, K. E. (1984) Paper No. 28, presented at the 126th National meeting of the Rubber Division, ACS, Denver, CO, 23–26 October.

Index

Page numbers appearing in **bold** refer to figures and page numbers appearing in *italic* refer to tables.

GPSR Compliance

The European Union's (EU) General Product Safety Regulation (GPSR) is a set of rules that requires consumer products to be safe and our obligations to ensure this.

If you have any concerns about our products, you can contact us on ProductSafety@springernature.com

In case Publisher is established outside the EU, the EU authorized representative is:

Springer Nature Customer Service Center GmbH
Europaplatz 3
69115 Heidelberg, Germany

Batch number: 10761599

Printed by Printforce, the Netherlands